Estimation In Linear Models

T. O. LEWIS
Associate Professor of Mathematics and Statistics
Texas Tech University, Lubbock, Texas

P. L. ODELL
Professor of Mathematics and Statistics
Texas Tech University, Lubbock, Texas

PRENTICE-HALL, INC. Englewood Cliffs, New Jersey

© 1971 by
PRENTICE-HALL, INC.
Englewood Cliffs, New Jersey

All rights reserved. No part of this book may be reproduced in any way, or by any means, without permission in writting from the publisher.

Current printing (last digit):
10 9 8 7 6 5 4 3 2 1

13-289967-1
Library of Congress Catalog Card Number: 74-149973
Printed in the United States of America

To **Mona** and **Norma**

Preface

This book is written for the scientist or engineer with a basic knowledge of elementary calculus and linear algebra. The intention of the presentation is not to give an axiomatic development of the theory of linear estimation but to make use of definitions and theorems to avoid inconsistences. Furthermore, the book is written so that it is essentially self-contained.

In Chapter 1, a brief discussion of vector spaces and matrices is developed. This includes the notion of generalized matrix inverses and their application to systems of linear equations.

Chapter 2 is concerned with statistical concepts, that is, it gives a brief development of probability spaces, random variables, and related topics. Also, the concept of an estimator is introduced, along with some desirable properties that it should possess.

In Chapters 3 and 4 a more extensive discussion of estimators is developed, and, for the most part, the discussion is restricted to linear estimators using discrete random sampling. In Chapter 3, the linear estimators of non-random parameters and random parameters are both considered; whereas, in Chapter 4 a linear recursive estimator is considered.

Chapters 5 and 7 are concerned with linear unbiased estimators using continuous data. Estimators of both the non-random and the random parameter are given.

In Chapter 6, the parameter to be estimated in the linear model is restricted to a specified region. If the restriction is a linear equality constraint, then conventional techniques are used; however, if the restriction is a linear inequality constraint, then quadratic programming is used.

The last four chapters are concerned with special topics, that is, in Chapter 8, if the covariance matrix of the random error vector in the linear model is unknown, then the advantages and disadvantages of well-known linear estimators are discussed. Chapter 9 deals with selecting an optimal design matrix in the linear model so as to minimize the variance of the Gauss-Markov estimators of the parameter vector. In Chapter 10 sufficient statistics and matrix lower bounds for the covariance matrix are discussed. Finally, in Chapter 11, least squares estimators in non-linear models are considered. Several well-known techniques are presented.

The notation used in this book is such that definitions, lemmas, and theorems are numbered according to chapter and section. Random variables will be denoted for the most part by capital letters and their values will be denoted by the corresponding lower case letter. Scalars and vectors will be denoted by lower case letters; whereas matrices will be indicated by capital letters.

Finally, the authors would like to express their gratitude to Patty Baker, Linda Woolfolk, Jennifer Birdwell, Kathy Dobbs, and Vicky Zwiacher for their time spent in typing the manuscript for this book.

T. O. LEWIS
P. L. ODELL

Contents

CHAPTER 1 Mathematical Concepts   **1**

    1-1 Vector Spaces, 1
    1-2 Matrices, 3
    1-3 The Generalized Inverse of an Arbitrary Matrix, 6
    1-4 Quadratic Forms, 12
    1-5 The Crout Factorization, 16
    1-6 Derivatives of Determinants and Matrices, 17
    1-7 A Matrix Variational Notation, 18
    1-8 Exercises, 21

CHAPTER 2 Statistical Concepts   **23**

    2-1 Introductory Comments, 23
    2-2 Random Variables and Density Functions, 23
    2-3 Expectations and Moments, 29
    2-4 Estimators, 35
    2-5 Exercises, 44

CHAPTER 3 Best Linear Estimation   **47**

    3-1 Introductory Remarks, 47
    3-2 The Linear Model, 50
    3-3 The Classical Form of the Gauss-Markov Theorem, 52
    3-4 Comparison of Least Squares and Minimum Variance Estimators, 59
    3-5 The Recursive Form of the Estimator, 60

3-6 The Gauss-Markov Theorem When the Parameter Vector is Random, 62
3-7 On Estimating a Subvector of X, 66
3-8 The Orthogonality Principle, 67
3-9 On Combining Estimators and Observations, 69
3-10 Exercises, 72

CHAPTER 4 Best Linear Recursive Estimation 75

4-1 Introductory Remarks, 75
4-2 A Useful Generalization, 78
4-3 An Example, 82
4-4 Exercises, 83

CHAPTER 5 Best Linear Estimation Using Continuous Data 85

5-1 Introduction, 85
5-2 A Best Linear Unbiased Estimator, 86
5-3 A Useful Formulation When X(T) is Random, 90
5-4 Relations Between Discrete and Continuous Models, 94
5-5 An Example, 96
5-6 Exercises, 97

CHAPTER 6 On Linear Estimation with Constraints 99

6-1 Introductory Remarks, 99
6-2 Estimation with Non-Random Linear Constraints, 100
6-3 Linear Constraints with Additive Random Components, 102
6-4 Linear Estimation with Inequality Constraints, 103
6-5 Exercises, 117

CHAPTER 7 Estimating a Stochastic Process in a Dynamic Model Using Continuous Data 119

7-1 Introductory Remarks, 119
7-2 The Dynamic Linear Model, 120
7-3 Concluding Remarks, 133
7-4 Exercises, 133

CHAPTER 8 Concerning Best Estimators Using an Estimated Covariance Matrix 135

    8-1 Introduction, 135
    8-2 Pertinent Matrix Identities, 136
    8-3 Covariance Adjustment, 139
    8-4 An Arbitrary Covariance Matrix V., 140
    8-5 Concluding Remarks, 142
    8-6 Exercises, 143

CHAPTER 9 On Selecting an Optimal Design Matrix 145

    9-1 Introduction, 145
    9-2 Rectangular and Ellipsoidal Regions, 151
    9-3 Exercises, 154

CHAPTER 10 Related Topics in Parametric Estimation 157

    10-1 Introductory Remarks, 157
    10-2 The Matrix Lower Bound, 157
    10-3 An Application, 160
    10-4 A Sufficient Statistic for a Parameter, 161
    10-5 Exercises, 166

CHAPTER 11 Least Square Estimates in a Non-Linear Model 169

    11-1 Introduction, 169
    11-2 Several Techniques for Solution of Non-Linear Least Square Problems, 169
    11-3 A Numerical Example, 173
    11-4 Exercises, 176

References   177

Index   191

Estimation
In Linear Models

CHAPTER 1

Mathematical Concepts

1-1 Vector Spaces

To establish a language and a notation that will make the discussion self-contained, several definitions and well-known facts concerning vector spaces and matrices are listed. It is not our intention to present an axiomatic development of the theory of linear estimation but to avoid inconsistency through use of the formalism of definitions and theorems.

Definition 1-1.1. *A vector space* V *is a collection of objects* x, y, z, ... *called vectors such that*

1. *If* x *and* y *are vectors in* V, *there is a unique vector* x + y *in* V *called the sum of* x *and* y.
2. *If* x, y, z *are vectors in* V, *and* α *and* β *are complex numbers, then the following is true (the symbol* $\in$ *should be read "an element of").*
   a. *There exists a unique vector* w $\in$ V *such that* $\alpha x = x\alpha = w$.
   b. *There exists a unique vector* k $\in$ V *such that* $\alpha\beta(x) = \alpha(\beta x) = k$.
   c. *Multiplication by scalars is distributive with respect to vector addition,* $\alpha(x + y) = \alpha x + \alpha y$.
   d. *Multiplication by vectors is distributive with respect to scalar addition,* $(\alpha + \beta)x = \alpha x + \beta x$.
   e. *Addition is commutative,* $x + y = y + x$.
   f. *Addition is associative,*

$$x + (y + z) = (x + y) + z.$$

3. $1x = x$ *for every vector* $x \in V$.
4. *For every* $x \in V$, *there exist vectors* 0 *and* $-x$, *satisfying*

$$x + 0 = 0 + x = x$$

$$x + (-x) = 0.$$

For convenience we write $x + (-y)$ *as* $x - y$.

Definition 1-1.2. *The vector space* V *is an inner-product space, also called a unitary space, if there is a uniquely defined complex number* $\langle x, y \rangle$ *called the "inner product" of the vectors* x *and* y, *which satisfies*

1. $\langle x, y \rangle = \langle \bar{y}, \bar{x} \rangle$
2. $\langle \alpha x, y \rangle = \bar{\alpha} \langle x, y \rangle$
3. $\langle x, x \rangle \geq 0$
4. $\langle x + y, z \rangle = \langle x, z \rangle + \langle y, z \rangle$
5. $\langle x, y + z \rangle = \langle x, y \rangle + \langle x, z \rangle$
6. $\langle x, x \rangle = 0$ *if and only if* $x = 0$.

The symbol $\overline{(\cdot)}$ *denotes the complex conjugate of* $(\cdot)$.

It is often convenient for one to think of a vector as a column of ordered *n*-tuples, such as

$$y = \begin{bmatrix} y_1 \\ y_2 \\ \vdots \\ y_n \end{bmatrix} = (y_1, y_2, \ldots, y_n)^T$$

and

$$y^* = (\bar{y}_1, \bar{y}_2, \ldots, \bar{y}_n),$$

where $(\cdot)^T$ denotes the transpose of $(\cdot)$ and $(\cdot)^*$ denotes the conjugate transpose of $(\cdot)$. We restrict our attention in this book almost entirely to the vector space of the set of all *n*-tuples whose elements are from the complex-number field. We call this space the *complex Euclidean n space* and denote it by the symbol C_n. If one restricts the elements of the vectors to the *real numbers*, a subfield of the field of complex numbers, then the space of *n*-tuples will be called simply the *Euclidean n-space*, denoted by E_n. An illustration of an inner-product space that will be used frequently throughout this book is the set C_n, such that if x, y are in C_n, then

$$x + y = (x_1 + y_1, x_2 + y_2, \ldots, x_n + y_n)^T$$

$$\alpha y = (\alpha x_1, \alpha x_2, \ldots, \alpha x_n),$$

where α is any complex scalar and

$$\langle x, y \rangle = \sum_{i=1}^{n} \bar{x}_i y_i.$$

We shall usually use x^*y for the "inner product."

Definition 1-1.3. *A set $\{x, y, \ldots, z\}$ of m distinct vectors from a vector space V are said to be linearly independent if there exists no set $\{\alpha, \beta, \ldots, \gamma\}$ of m scalars except $\alpha = \beta = \ldots = \gamma = 0$ such that $\alpha x + \beta y + \ldots + \gamma z = 0$.*

Definition 1-1.4. *Two vectors x and y are said to be orthogonal if $x^*y = 0$.*

Definition 1-1.5. *Two vector spaces V and U are said to be orthogonal if for every vector x in V and every vector y in U, x and y are orthogonal.*

The number of definitions listed here is minimal and it is suggested for those readers whose background in the theory of vector spaces and matrices is marginal, to refer as needed to one of the many good elementary texts [82, 88, 112, 144] on these topics.

1-2 Matrices

The theory of linear estimation as developed in this book requires some knowledge of the theory and manipulation of matrices defined on the complex field.

All matrices will be designated by a capital English letter; the notation that

$$A = \{a_{ij}\}$$

means that A is an $m \times n$ matrix having a_{ij} as the element of the ith row and jth column. The letters m and n denote the number of rows and columns, respectively. Clearly, if $m = n$, the matrix A is square. If A denotes a matrix, then A^T and A^* denote the transpose and the conjugate transpose of the matrix A. If the elements a_{ij} of the matrix A are real numbers, then $A^T = A^*$. The symbol $|A|$ will denote the determinant of A when A is square, and A^{-1} will denote the inverse of A when it exists. The symbols I and 0 will denote the identity matrix and the null matrix, respectively. The null matrix is a matrix whose elements are all zeros.

Every effort has been made to restrict the use of lowercase English letters to vectors, that is, $n \times 1$ matrices. Given any vector x, the symbol

$$\|x\| = (x^*x)^{1/2}$$

denotes the *norm* (a scalar) of x.

In the definitions that follow, A is an $m \times n$ matrix defined on the complex field.

Definition 1-2.1. *The rank of an* $m \times n$ *matrix* A *is the number of linearly independent rows, when the rows of* A *are considered as* $n \times 1$ *vectors (see Definition 1-1.3). We denote the rank of* A *by the symbol* $r(A)$.

It should be noted that if one counts the number of linearly independent columns of the $m \times n$ matrix A, this number equals the rank of A.

Definition 1-2.2. *The column or range space* $R(A)$ *of an* $m \times n$ *matrix is the set of all vectors* Ax *as* x *ranges over the space* C_n. *The row space of a matrix* A *is the set of all vectors* $x^T A$ *as* x *ranges over the space* C_m.

If one lets $y = Ax$, where $A = (a_{ij})$ is an $m \times n$ matrix, then $R(A)$ is a subset of C_m, where the elements of y are defined by

$$y_i = \sum_{j=1}^{n} a_{ij} x_j, \qquad i = 1, 2, \ldots, m.$$

Definition 1-2.3. *The null space* $N(A)$ *of a matrix* A *is the set of all vectors* x *in* C_n *such that* $Ax = 0$ (*the* $n \times 1$ *null vector*). *The row null space of* A *is similarly defined.*

A set of interest in our discussion is the orthogonal complement of $N(A)$ denoted by the symbol $N(A)^\perp$, the set of vectors orthogonal to every vector in $N(A)$. It can be shown that if A is an $m \times n$ matrix of rank r, then the maximum number of linear independent vectors in $R(A)$ is r.

THEOREM 1-2.1. $N(A)^\perp = R(A^*)$.

Proof: Let A be an $m \times n$ matrix of rank r; then the maximum number of linear independent vectors in $R(A^*)$ is r. The maximum number of linear independent vectors in $N(A)$ is $n - r$. Now $R(A^*) \cap N(A) = \theta$, where θ is the null space, not required if to much spacing involved, suppose there exists $x \neq 0$ such that $x \in R(A^*) \cap N(A)$. Then $x \in N(A)$ and $x \in R(A^*)$, which implies there exists y such that $A^*y = x$. Thus $AA^*y = 0$, which implies

$A^*y = 0$. This is impossible. Hence $R(A^*) \cap N(A) = \theta$. Thus $R(A^*) \oplus N(A) = E_n$.

Let $x \neq 0$ and $x \in R(A^*)$. Then there exists a y such that $A^*y = x$. Let $x_0 \in N(A)$, then $x_0^* x = x_0^* A^* y = (Ax_0)^* y = 0$, which implies $x \in N(A)^\perp$. Thus $R(A^*) \subset N(A)$. Since the maximum number of linear independent vectors in $R(A^*)$ is equal to the maximum number of linear independent vectors in $N(A)^\perp$, it follows that $R(A^*) = N(A)^\perp$.

Before stating Definition 1-2.4, let us digress to the following illustration. Suppose we are given the system of linear equations

$$Ax = y, \qquad (1\text{-}1)$$

and we desire a solution x_0 of this system. Further, suppose there exists a matrix B such that

$$AB = I, \qquad (1\text{-}2)$$

then a solution of (1-1) is By denoted by x_0; that is,

$$Ax_0 = A(By) = ABy = y.$$

The matrix B is the right inverse of A. This brief discussion motivates a concept of an inverse of a matrix.

Definition 1-2.4. *Let* A *be a square matrix whose rank equals its dimension; then there exists a matrix* B *such that*

$$AB = BA = I.$$

The matrix B *is usually denoted by* B $= $ A^{-1}.

It is important to note, however, that in order to solve (1-1), the matrix B need only be a right inverse of A.

Definition 1-2.5. *If* A *is an* m $\times$ n *matrix, where* r(A) $= $ m $\leq$ n, *then* A *has a right inverse defined by* $A^T(AA^T)^{-1} + [I - A^T(AA^T)^{-1}A]$. *The symbol* r(A) *denotes the rank of the matrix* A.

The general solution of (1-1) is then

$$x_0 = A^T(AA^T)^{-1}y + [I - A^T(AA^T)^{-1}A]z,$$

where z is arbitrary $n \times 1$ vector.

A far more interesting and practical problem is to solve (1-1) when A is $m \times n$ and $r(A) = n \leq m$. There are many cases, when y is not in $R(A)$, wherein no solution for x exists and it becomes necessary to obtain an approximate solution of (1-1). When one proposes an approximate solution, he introduces an element of arbitrariness. That is, he must define what he means by a "good" approximation. A natural extension to the concept of an inverse of a matrix introduces a criterion for a best approximate solution that is not only intuitively reasonable, but also allows one to unify the theory of linear estimation.

1-3 The Generalized Inverse of an Arbitrary Matrix

A convenient way to introduce the concept of the pseudoinverse of a matrix is to introduce the following sequence of definitions of generalized matrix inverses [134, 142, 165].

Definition 1-3.1. The matrix X is a generalized inverse for A if and only if $AXA = A$

Definition 1-3.2. The matrix X is a reflexive generalized inverse for A if and only if $AXA = A$ and $XAX = X$.

Definition 1-3.3. The matrix X is a normalized generalized inverse of A if and only if $AXA = A$, $XAX = X$, and AX is hermitian; that is, $(AX)^* = AX$.

Definition 1-3.4. The matrix X is a pseudoinverse for A if and only if $AXA = A$, $XAX = X$, AX is hermitian, and XA is hermitian.

If A is an $m \times n$ matrix and the rank of A is r, then there exist nonsingular matrices P and Q such that

$$PAQ = \begin{bmatrix} I_r & \phi \\ \phi & \phi \end{bmatrix} = J_r,$$

where I_r denotes an $r \times r$ identity matrix. We say that A is equivalent to J_r.

THEOREM 1-3.1. Let A be equivalent to J_r, that is, $PAQ = J_r$ and

$$X = Q \begin{bmatrix} Z & U \\ V & W \end{bmatrix} P.$$

Then X is a generalized inverse of A if and only if $Z = I_r$.

Proof: Since $PAQ = J_r$ and P and Q are nonsingular, then $A = P^{-1}J_rQ^{-1}$. Let $Z = I_r$, then by matrix multiplication $AXA = A$. Also, if X is a generalized inverse of A,

$$AXA = P^{-1}J_rQ^{-1}Q\begin{bmatrix} Z & U \\ V & W \end{bmatrix}PP^{-1}J_rQ^{-1} = P^{-1}\begin{bmatrix} Z & \phi \\ \phi & \phi \end{bmatrix}Q^{-1}.$$

We solve for Z in the matrix equation

$$P^{-1}\begin{bmatrix} Z & \phi \\ \phi & \phi \end{bmatrix}Q^{-1} = P^{-1}\begin{bmatrix} I_r & \phi \\ \phi & \phi \end{bmatrix}Q^{-1}$$

and find that $Z = I_r$, the desired result.

COROLLARY 1-3.1. *If the matrix* A *is hermitian, then* $X = PJ_rP^*$ *is a hermitian generalized inverse for* A, *where* P *is the nonsingular matrix such that* $PAP^* = J_r$.

Proof: From elementary matrix theory we know that if A is hermitian, there exists a nonsingular matrix P such that $PAP^* = J_r$. Letting P^* be Q in Theorem 1-3.1 gives $X = PJ_rP^*$. Finally, we note that $X^* = X$.

LEMMA. $A^*A = \phi$ *implies* $A = \phi$.

Proof: Let the ith column of $A = \{a_{ij}\}$ be A_i, and $C = \{c_{ij}\} = A^*A$. The jth diagonal element of C, say, $c_{jj} = A_j^*A_j = \sum_{i=1}^{n}\bar{a}_{ij}a_{ij}$, is a sum of positive quantities. But by hypothesis, $c_{jj} = 0$ for every j, which implies that every element of A_j is zero; that is, $A = \phi$.

THEOREM 1-3.2. $BAA^* = CAA^*$ *implies* $BA = CA$.

Proof: Let $BAA^* - CAA^* = \phi$; then $\phi = (BAA^* - CAA^*)(B - C)^* = (BA - CA)(BA - CA)^*$. By the lemma, $BA - CA = \phi$ or $BA = CA$.

Selecting any generalized inverses of AA^* and A^*A, say, $(AA^*)^g$ and $(A^*A)^g$, respectively, and using Theorem 1-3.2, one can easily establish the following identities:

$$A = AA^*(AA^*)^g A$$

$$A = A(A^*A)^g A^*A$$

$$A = AA^*[(AA^*)^g]^*A$$

$$A = A[(A^*A)^g]^*A^*A.$$

THEOREM 1-3.3. *Let* A^g, A^r, A^n, *and* A^+ *denote a generalized, reflexive, normalized, and the pseudoinverse of the matrix* A, *respectively; then*

$$A^r = A^g A A^g \tag{1-3}$$

$$A^n = (A^*A)^g A^* \tag{1-4}$$

$$A^+ = A^*(AA^*)^g A(A^*A)^g A^* \tag{1-5}$$

Proof: The proof follows by observing that (1-3)–(1-5) satisfy the definitions of A^r, A^n, and A^+, respectively.

THEOREM 1-3.4. *The pseudoinverse of the matrix* A *is unique.*

Proof: Let X be a pseudoinverse of the matrix A; then it can be shown that the following relations hold:

$$XX^*A^* = X \tag{1-6}$$

$$XAA^* = A^*. \tag{1-7}$$

Now suppose that Y is any other pseudoinverse of A; then it also follows that

$$A^*Y^*Y = Y \tag{1-8}$$

$$A^*AY = A^*. \tag{1-9}$$

Thus it follows that

$$X = XX^*A^* = XX^*A^*AY = XAY = XAA^*Y^*Y = A^*Y^*Y = Y.$$

Hence $X = Y$, which implies A^+, is unique.

It is important to note that if the set of all generalized inverses, reflexive inverses, normalized inverses, and pseudoinverses is denoted by $\{A^g\}$, $\{A^r\}$, $\{A^n\}$, and $\{A^+\}$, respectively, then

$$\{A^+\} \subset \{A^n\} \subset \{A^r\} \subset \{A^g\},$$

where $U \subset V$ implies that the set U is contained in the set V.

We now give a definition of a *norm* of a matrix A, which we denote by $\|A\|$.

Definition 1-3.5. *The norm of a matrix* A *is the square root of the sum of squares of the moduli of the elements of* A; *that is,* $\|A\| = (\text{trace})^{1/2} A^*A$, *where the trace of a square matrix* A *is simply the sum of its diagonal elements.*

Definition 1-3.6. *The matrix* X_0 *is a best approximate solution of the equation* $f(X) = G$ *if for all X, either*

1. $\|f(X) - G\| > \|f(X_0) - G\|$, or

2. $\|f(X) - G\| = \|f(X_0) - G\|$, and

$$\|X\| \geq \|X_0\|.$$

THEOREM 1-3.5. A^+B *is the unique best approximate solution of the equation* $AX = B$, *where A, X, and B are matrices.*

Proof: The identity

$$\|AP + (I - AA^+)Q\|^2 = \|AP\|^2 + \|(I - AA^+)Q\|^2 \tag{1-10}$$

holds since

$$\{AP + (I - AA^+)Q\}^*\{AP + (I - AA^+)Q\}$$
$$= (AP)^*AP + \{(I - AA^+)Q\}^*(I - AA^+)Q.$$

Thus in particular

$$\|AX - B\|^2 = \|A(X - A^+B) + (AA^+ - I)B\|^2$$
$$= \|AX - AA^+B\|^2 + \|AA^+B - B\|^2,$$

which implies

$$\|AX - B\| \geq \|A(A^+B) - B\|$$

with equality only when $AX = AA^+B$. Replacing A by A^+ in (1-10) and using the fact that $A^{++} = A$, one can deduce

$$\|A^+B + (I - A^+A)X\|^2 = \|A^+B\|^2 + \|(I - A^+A)X\|^2.$$

Thus, if $AX = AA^+B$ and since $A^+AA^+ = A^+$, it follows that

$$\|X\|^2 = \|A^+B\|^2 + \|X - A^+B\|^2 \geq \|A^+B\|^2,$$

or finally,

$$\|X\| \geq \|X_0\|.$$

COROLLARY 1-3.2. *The best approximate solution of* $AX = I$ *is* A^+. A very useful result is summarized in Theorem 1-3.6.

THEOREM 1-3.6. *A necessary and sufficient condition for the equation*

$$AXB = C,$$

where A, X, B, *and* C *are matrices, to have a solution is*

$$AA^+CB^+B = C,$$

in which case the general solution is

$$X = A^+CB^+ + Y - A^+AYBB^+,$$

where Y *is an arbitrary matrix.*

Proof: Suppose that X satisfies $AXB = C$. Then

$$C = AXB = AA^+AXBB^+B = AA^+CB^+B.$$

Conversely, if $C = AA^+CB^+B$, then A^+CB^+ is a particular solution of $AXB = C$.

For the general solution one must solve $AXB = 0$. Since A^+ and B^+ always exist, then it follows immediately that $Y - AA^+YBB^+$ is a solution. Furthermore, if $AXB = 0$, then $X = X - A^+AXBB^+$, which implies the general solution of $AXB = C$ is $X = A^+CB^+ + Y - A^+AYBB^+$ whenever $AA^+CB^+B = C$. Thus the theorem is proved.

It is important to note that the only requirement of A^+ for Theorem 1-3.6. is that $AA^+A = A$; hence the theorem will hold if A^+ is replaced by A^g, A^r, or A^n.

COROLLARY 1-3.3. *The general solution of the vector equation* $Ax = b$ *is*

$$x = A^+b + (I - A^+A)y,$$

where y *is arbitrary, provided that the equation has a solution.*

It should be noted that $x = A^+b$ is the usual least square solution for the matrix equation $Ax = b$. The least square solution is the *best approximate solution*.

We list the following useful properties of the pseudoinverse without proof, most of which follow directly from the definition and the uniqueness of the pseudoinverse.

1. If $D = (d_{ij})$ is a square ($m = n$) and diagonal ($d_{ij} = 0$ for $i \neq j$) then $D^+ = (d_{ij}^+)$ is defined by $d_{ij}^+ = 0$ for $i \neq j$, $d_{ii}^+ = 0$ if $d_{ii} = 0$, and $d_{ii}^+ = d_{ii}^{-1}$ if $d_{ii} \neq 0$.

2. If $A^*A = PDP^*$, where $PP^* = P^*P = I$, and D is diagonal, then $A^+ = PD^+P^*A^*$.
3. If $A = BC$, where the columns of B are linearly independent and the rows of C are linearly independent, then

$$A^+ = C^*(CC^*)^{-1}(B^*B)^{-1}B^*.$$

4. $A^+ = (A^*A)^{-1}A^*$, if the columns of A are linearly independent.
5. $A^+ = A^*(AA^*)^{-1}$, if the rows of A are linearly independent.
6. $A^+ = A^{-1}$, if A is square and nonsingular.
7. $(A^+)^+ = A$.
8. $(A^*)^+ = (A^+)^* \equiv A^{+*} \equiv A^{*+}$.
9. $A^+AA^* = A^*$.
10. $A^*AA^+ = A^*$.
11. $AA^+A^{+*} = A^{+*}$.
12. $A^{+*}A^+A = A^{+*}$.
13. $A^{*+}A^*A = A$.
14. $AA^*A^{*+} = A$.
15. $A^*A^{+*}A^+ = A^+$.
16. $A^+A^{+*}A^* = A^+$.
17. The row spaces of A^+ and A^* are identical; that is, the rows of A^+ are in the row space of A^* and the rows of A^* are in the row space of A^+.
18. The column spaces of A^+ and A^* are identical.
19. A, A^+, and A^* all have the same rank.
20. $(AA^*)^+ = A^{+*}A^+$.
21. $(AA^*)^+(AA^*) = AA^+$.
22. If A^+ commutes with some power of A and λ is any nonzero eigenvalue of A corresponding to the eigenvector x, then λ^{-1} is an eigenvalue of A^+ corresponding to the eigenvector x (eigenvalues or characteristic roots and eigenvectors or characteristic vectors will be defined in Section 1-4).
23. If $\alpha \neq 0$ then $(\alpha A)^+ = \alpha^{-1}A^+$.
24. $0^+ = 0^T$.
25. A^+A, AA^+, $I - A^+A$, and $I - AA^+$ are hermitian idempotent. (A square matrix H is said to be idempotent if $HH = H$.)

THEOREM 1-3.7. *Let* A *and* B *be matrices with the product* AB *defined. Then*

$$(AB)^+ = B_1^+A_1^+,$$

where

$$AB = A_1B_1$$

$$B_1 = A^+AB$$

$$A_1 = AB_1B_1^+$$

Proof: The product AB can be written as

$$AB = AA^+AB = AB_1 = AB_1B_1^+B_1 = A_1B_1.$$

Let $Y = AB = A_1B_1$ and let $X = B_1^+A_1^+$. Then it is only necessary to show that Y and X satisfy the four defining equations in Definition 1-3.4 for the pseudo-inverse. From the definition of A_1 we have $A_1B_1B_1^+ = AB_1B_1^+B_1B_1^+ = A_1$. Now $YX = A_1B_1B_1^+A_1^+ = A_1A_1^+$ is hermitian. Also, $YXY = A_1B_1B_1^+A_1^+A_1B_1 = A_1A_1^+A_1B_1 = A_1B_1 = Y$ and $XYX = B_1^+A_1^+(A_1B_1B_1^+)A_1^+ = B_1^+A_1^+A_1A_1^+ = B_1^+A_1^+ = X$. To show that XY is hermitian, we observe first, using the definitions of A_1 and B_1, that

$$A^+A_1 = A^+AB_1B_1^+ = A^+A(A^+AB)B_1^+ = A^+ABB_1^+ = B_1B_1^+.$$

Also, since $A_1^+A_1B_1B_1^+ = A_1^+A_1$, with both $A_1^+A_1$ and $B_1B_1^+$ hermitian, $B_1B_1^+A_1^+A_1 = A_1^+A_1$. Substituting A^+A_1 for $B_1B_1^+$ gives $A_1^+A_1 = A^+A_1A_1^+A_1 = A^+A_1$, and so $A_1^+A_1 = B_1B_1^+$.

From this it now follows that $XY = B_1^+A_1^+A_1B_1 = B_1^+B_1B_1^+B_1 = B_1^+B_1$ is hermitian. Since it has been shown that Y and X satisfy the defining equations for the pseudoinverse, $X = Y^+$. But $X = B_1^+A_1^+$.

It is of interest to show that $(AB)^+ = B^+A^+$ under certain conditions imposed on A and B. Theorem 1-3.8 was obtained by T. N. E. Greville [80].

THEOREM 1-3.8. *A necessary condition for* $(AB)^+ = B^+A^+$ *is that* A^+A *and* BB^+ *commute.*

Note that if both A^{-1} and B^{-1} exist, then $(AB)^+ = (AB)^{-1} = B^{-1}A^{-1}$.

THEOREM 1-3.9. *Let* A *be a* p $\times$ p *nonsingular matrix; then* $A^{-1} = \{cof\ a_{ij}\}^T/|A|$, *where* $\{cof\ a_{ij}\}$ *is the matrix of i-jth cofactor.*

1-4 Quadratic Forms

Let A be an $n \times n$ hermitian matrix and x be an $n \times 1$ vector; then

$$Q = x^*Ax$$

is a real number. The quantity Q is called a quadratic form.

Sec. 1-4 Quadratic Forms

Definition 1-4.1. *If for every* x *except the null vector* $Q > 0$, *then the matrix* A *is said to be a positive definite matrix.*

Definition 1-4.2. *If there exists at least one vector* $x \neq 0$ *such that* $Q = 0$ *and* $Q \geq 0$ *for all remaining vectors* x, *then* A *is said to be a positive semidefinite matrix.*

We list without proof several theorems that are useful.

THEOREM 1-4.0. *If* P *is a nonsingular matrix and* A *is positive definite (positive semidefinite), then* P*AP *is positive definite (positive semidefinite).*

THEOREM 1-4.1. *The hermitian matrix* A *is positive definite if and only if there exists a nonsingular matrix* P *such that* A = P*P.

THEOREM 1-4.2. *The determinant of a positive definite matrix and all its principal subdeterminants are positive.*

THEOREM 1-4.3. *If* A *is an* n × m *matrix of rank* m < n, *then* A*A *is positive definite and* AA* *is positive semidefinite.*

THEOREM 1-4.4. *If* A *is an* n × m *matrix of rank less than* m *or* n, *then* A*A *and* AA* *are each positive semidefinite.*

Definition 1-4.3. *A characteristic root or eigenvalue of an* n × n *matrix* A *is a scalar* λ *such that* $Ax = \lambda x$ *for some vector* $x \neq 0$. *The vector* x *is called the characteristic vector or eigenvector of the matrix* A.

THEOREM 1-4.5. *The number of nonzero eigenvalues of a hermitian matrix* A *is equal to the rank of* A.

THEOREM 1-4.6. *The eigenvalues of a hermitian matrix are real.*

THEOREM 1-4.7. *The eigenvalues of a positive semidefinite matrix are nonnegative.*

THEOREM 1-4.8. *For every hermitian matrix* A *there exists a unitary matrix* P *such that* P*AP = D, *where* D *is a diagonal matrix whose elements are the eigenvalues of* A.

It is sometimes advantageous to break a matrix into submatrices. This is called partitioning a matrix into submatrices. The following example will illustrate this procedure. Let A be an $n \times n$ matrix and write

$$A = \begin{bmatrix} A_{11} & A_{12} \\ A_{21} & A_{22} \end{bmatrix},$$

where A_{11} is $n_1 \times m_1$, A_{12} is $n_1 \times (n - m_1)$, A_{21} is $(n - n_1) \times m_1$, and A_{22} is $(n - n_1) \times (n - m_1)$.

The product AB of two matrices can be written symbolically when A and B are partitioned compatibly into submatrices. The multiplication proceeds as though the submatrices were single elements of the matrix. However, the dimensions of the matrices and of the submatrices must be such that they will multiply.

THEOREM 1-4.9. *If* A *is a positive definite hermitian matrix such that*

$$A = \begin{bmatrix} A_{11} & A_{12} \\ A_{21} & A_{22} \end{bmatrix}$$

and if B *is the inverse of* A *such that*

$$B = \begin{bmatrix} B_{11} & B_{12} \\ B_{21} & B_{22} \end{bmatrix},$$

then

$$B_{11} = [A_{11} - A_{12}A_{22}^{-1}A_{21}]^{-1}$$

$$B_{22} = [A_{22} - A_{21}A_{11}^{-1}A_{12}]^{-1}$$

$$B_{12} = -A_{11}^{-1}A_{12}B_{22}$$

$$B_{21} = -A_{22}^{-1}A_{21}B_{11}.$$

Also, we can write

$$B_{11} = A_{11}^{-1} + A_{11}^{-1}A_{12}[A_{22} - A_{21}A_{11}^{-1}A_{12}]^{-1}A_{21}A_{11}^{-1}$$

$$B_{22} = A_{22}^{-1} + A_{22}^{-1}A_{21}[A_{11} - A_{12}A_{22}^{-1}A_{21}]^{-1}A_{12}A_{22}^{-1}.$$

The latter formulas for B_{11} and B_{22} are sometimes called the "inside-out rule," which is proved formally in Theorem 1-4.10 in a slightly modified form.

THEOREM 1-4.10. *If* $P_1 = P_0 + A^*A$, *where* P_0 *is an* (n × n) *positive definite matrix and* A *is any* r × n *matrix, then*

$$P_1^{-1} = P_0^{-1} - P_0^{-1}A^*(AP_0^{-1}A^* + I)^{-1}AP_0^{-1} \qquad (1\text{-}11)$$

Proof: Since P_0 is positive definite, then P_0^{-1} is positive definite. Hence it follows $(AP_0^{-1}A^* + I)$ is positive definite, which implies $(AP_0^{-1}A^* + I)^{-1}$ exists. Therefore,

$$P_1^{-1}P_1 = I + P_0^{-1}A^*A - P_0^{-1}A^*(AP_0^{-1}A^* + I)(AP_0^{-1}A^* + I)^{-1}A$$

$$= I.$$

The above inversion formula has been used extensively in the sequential estimation theory for updating estimates as more samples are observed [50, 85]. It is of interest to know when a formula similar to (1-11) holds for pseudoinverses, since in some applications P_0 may be hermitian positive semidefinite [118].

THEOREM 1-4.11. *If P_0 is a positive semidefinite (hermitian)* m × m *matrix and* A *is an* r × m *matrix with* $P_1 = P_0 + A^*A$, *then*

$$P_1^+ = P_0^+ - P_0^+A^*(AP^+A^* + I)^{-1}AP^+$$

if, and only if, the null space of A *contains the null space of* P_0.

THEOREM 1-4.12. *If* A *is a square matrix such that*

$$A = \begin{bmatrix} A_{11} & A_{12} \\ A_{21} & A_{22} \end{bmatrix},$$

where A_{11} *and* A_{22} *are square matrices, and if* $A_{12} = 0$ *or* $A_{21} = 0$, *then* $|A| = |A_{11}||A_{22}|$.

THEOREM 1-4.13. *Let* A *be as in Theorem* 1-4.12. *If* A_{22} *is nonsingular, then*

$$|A| = |A_{22}||A_{11} - A_{12}A_{22}^{-1}A_{21}|.$$

Proof: Let

$$X = \begin{bmatrix} I & 0 \\ -A_{22}^{-1}A_{21} & A_{22}^{-1} \end{bmatrix},$$

then

$$|X| = |A_{22}^{-1}|$$

by Theorem 1-4.12. Since $|A_{22}^{-1}| = 1/|A_{22}|$ we can write

$$|A| = |A_{22}||A||A_{22}^{-1}|.$$

By replacing A by its partition and $|A_{22}^{-1}|$ by $|X|$ it follows that

$$|A| = |A_{22}| \begin{vmatrix} A_{11} & A_{12} \\ A_{21} & A_{22} \end{vmatrix} \begin{vmatrix} I & 0 \\ -A_{22}^{-1}A_{21} & A_{22}^{-1} \end{vmatrix}.$$

Since for conformable square matrices $|AB| = |A||B|$, it follows that

$$|A| = |A_{22}||AX|$$
$$= |A_{22}||A_{11} - A_{12}A_{22}^{-1}A_{21}|.$$

1-5 The Crout Factorization

Let $P = (p_{ij})$, $i, j = 1, 2, \ldots, n$, be positive definite, real, symmetric matrix. A slight modification of Theorem 1-4.2. allows one to find a factorization of P such that

$$P = TT^T, \tag{1-12}$$

where T is a lower triangular square matrix, with positive elements on the main diagonal. If the existence of the factorization (1-12) is given, then it is easy to show how to compute the components t_{ij} of T in the order

$$ij = 11, 21, \ldots, n1; \; 22, 32, \ldots, n2; \; \ldots, nn.$$

Since $t_{ij} = 0$ for $j > i$ then (1-12) allows one to write

$$p_{ij} = \sum_{k=1}^{j} t_{ik} t_{jk}. \tag{1-13}$$

First we compute

$$t_{11} = (p_{11})^{1/2}. \tag{1-14}$$

The other elements in the first column are

$$t_{i1} = t_{11}^{-1} p_{i1}, \qquad i = 2, 3, \ldots, n. \tag{1-15}$$

If the preceding columns $k < j$ have been computed, we compute the diagonal element

$$t_{jj} = \left(p_{jj} - \sum_{k=1}^{j-1} t_{jk}^2\right)^{1/2}. \tag{1-16}$$

If $j < n$, the elements below the diagonal are computed from the formula

$$t_{ij} = t_{jj}^{-1}\left(p_{ij} - \sum_{k=1}^{j-1} t_{ik} t_{jk}\right) \qquad i = j+1, \ldots, n. \tag{1-17}$$

1-6 Derivatives of Determinants and Matrices

It is useful to be able to compute the derivatives of determinants and matrices with respect to an element of the matrix. Hence we list and review several definitions and results relating to determinants and matrices.

Definition 1-6.1. *Let* $A = \{a_{ij}\}$ *be in an* $n \times n$ *matrix and* A_{ij} *be an* $(n-1) \times (n-1)$ *submatrix of* A *obtained by deleting row* i *and column* j *of* A. *The determinant* $|A_{ij}|$ *of the matrix* A_{ij} *is said to be the minor of* a_{ij} *and the cofactor of* a_{ij} *if* $(-1)^{i+j}|A_{ij}|$, *denoted by* $cof(a_{ij})$.

It is well known that

$$A = \sum_{i=1}^{n} a_{ij}(-1)^{i+j} A_{ij}, \tag{1-18}$$

the expansion by the *j*th column; and

$$A = \sum_{j=1}^{n} a_{ij}(-1)^{i+j} A_{ij}, \tag{1-19}$$

the expansion by the *i*th row.

THEOREM 1-6.1. *Let* A *be an* $n \times n$ *matrix. Then*

$$\frac{\partial |A|}{\partial a_{ij}} = (-1)^{i+j}|A_{ij}| = \mathrm{cof}(a_{ij}).$$

Proof: The theorem follows directly from the formula (1-18) or (1-19).

COROLLARY 1-6.1. *If* $A = A^T$, *then*

$$\frac{\partial |A|}{\partial a_{ij}} = \mathrm{cof}(a_{ij}), \quad i = j$$

and

$$\frac{\partial |A|}{\partial a_{ij}} = 2\,\mathrm{cof}(a_{ij}), \quad i \neq j.$$

Let the scalar $y = f(x_1, x_2, \ldots, x_p)$ be a function of the independent variables $x_1, x_2, \ldots, x_p$ such that all the first partial derivatives exist; then we mean by $\partial y/\partial x$ the $p \times 1$ vector whose *i*th element is $\partial f/\partial x_i$.

THEOREM 1-6.2. *Let* y = ax, *where* a *is* 1 × p, *and* x *is* p × 1; *then it is convenient to write the* p × 1 *vector of partial derivatives of* y *as* ∂y/∂x, *and*

$$\frac{\partial y}{\partial x} = \left\{\frac{\partial y}{\partial x_i}\right\} = a^T. \tag{1-20}$$

THEOREM 1-6.3. *Let* Q *be a quadratic form defined by*

$$Q = x^T A x; \tag{1-21}$$

then

$$\frac{\partial Q}{\partial x} = 2Ax. \tag{1-22}$$

Proof:

$$\frac{\partial Q}{\partial x} = \left\{\frac{\partial}{\partial x_i} \sum_{i,j}^{n} x_i a_{ij} x_j\right\}$$

$$= \left\{\frac{\partial}{\partial x_i}\left[\sum_{j=1}^{n} a_{jj} x_j^2 + 2 \sum_{j \neq i}^{n} a_{ij} x_i x_j\right]\right\}$$

$$= 2 \sum_{j=1}^{n} a_{jj} x_j + 2 \sum_{j \neq i}^{n} a_{ij} x_j$$

$$= 2\left\{\sum_{j=1}^{n} a_{ij} x_j\right\}$$

$$= 2Ax,$$

the desired result.

1-7 A Matrix Variational Notation

It is convenient at times, especially in determining maxima and minima of matrix-valued functions, to use a so-called *variational notation* for matrices [54]. One will note the relation to the more familiar usage of the notation in the calculus of variation.

Definition 1-7.1. *A positive definite matrix* A *is said to be smaller than a positive definite matrix* B *if* B − A *is positive semidefinite or positive definite.*

It is a fact that there exists a neighborhood N of every positive definite

Sec. 1-7 A Matrix Variational Notation

matrix A such that if

$$A + \delta A \in N$$

then $A + \delta A$ is also positive definite.

Definition 1-7.2. *The positive definite matrix-valued function* f *of the matrix* X *is said to possess a minimal if there exists an* X, *say,* X_0, *such that*

$$f(X) - f(X_0)$$

is positive semidefinite for all X *in an "arbitrarily small" neighborhood of* X_0.

Definition 1-7.3. *The first variation of the following matrix functions* AX, $X^T A$, $X^T AX$, *and* XAX^T *are, respectively,* $A\delta X$, $\delta X^T A$, $\delta X^T AX + X^T A \delta X$, *and* $\delta XAX^T + XA\delta X^T$. *The incremental differential operation* δX *is called a variation of* X.

THEOREM 1-7.1. *Let* $f(X) = XAX^T$ *be a positive definite matrix equation. A necessary condition for* $f(X)$ *to have a minimal is that* $\delta_X f$, *the first variation of* f *with respect to* X, *vanish.*

Proof: Consider $f(X)$ to be minimized for $X = X_0$; then

$$f(X_0 + \delta X) - f(X_0) = \delta X A X_0^T + X_0 A \delta X^T + \delta X A \delta X^T$$

must be at least positive semidefinite for every δX. We note that $\delta X A \delta X^T$ is positive semidefinite; hence we conclude that $A X_0^T = 0$, or if A is nonsingular (as is usually the case), $X_0 = 0$. Thus we can conclude that $\delta_X f \equiv 0$ is a necessary condition for $f(X)$ to be minimal.

The concept of Lagrangian multipliers is easily extended to the variational calculus for matrices.

THEOREM 1-7.2. *Let* $f(X) = XAX^T$ *be a matrix-valued function subject to the linear constraint*

$$XB = C, \tag{1-23}$$

where X *is a* p × n *matrix,* A *is a positive definite matrix,* B *is an* n × r *constant matrix whose rank is* r < n, *and* C *is a* p × r *constant matrix, then*

$$X_0 = C(B^T A^{-1} B)^{-1} B^T A^{-1}$$

is the minimizing matrix and the minimal value is $f(X_0) = C^T (B^T A^{-1} B)^{-1} C$.

Proof: Consider the Lagrange equation,

$$f(X) = XAX^T + [C - XB]\lambda + \lambda^T[C^T - B^T X^T],$$

where λ is the r × p Lagrange multiplier matrix. The first variation of f with respect to X is

$$\delta_x f = \delta X[AX^T - B\lambda] + [XA - \lambda^T B^T]\delta X^T.$$

A necessary condition for a minimal implies that

$$AX^T - B\lambda = 0$$

or

$$AX^T = B\lambda$$

$$X^T = A^{-1}B\lambda. \qquad (1\text{-}24)$$

Multiply both sides of this last quantity by B^T and, applying (1-23),

$$B^T X^T = B^T A^{-1} B\lambda$$

or

$$C^T = B^T A^{-1} B\lambda.$$

Solving for λ, we obtain

$$\lambda = (B^T A B)^{-1} C^T. \qquad (1\text{-}25)$$

We have selected the rank of B so that $B^T A B$ is nonsingular.
Substituting (1-25) into (1-24) one obtains

$$X^T = A^{-1} B (B^T A B)^{-1} C^T$$

or

$$X_0 = C(B^T A B)^{-1} B^T A^{-1},$$

the desired result.
We note that $X_0 B = C$, and

$$f(X_0) = C(B^T A B)^{-1} B^T A^{-1} A A^{-1} B (B^T A B)^{-1} C^T$$

$$f(X_0) = C(B^T A B)^{-1} C^T.$$

EXERCISES

1. A matrix norm and a vector norm are said to be consistent if $\|Ax\| \leq \|A\|\|x\|$ for any A and x. Show that the matrix and vector norm defined in this chapter are consistent.

2. Prove Theorems 1-4.1 through 1-4.5.

3. Prove Theorems 1-4.6 through 1-4.10, 1-4.12, and 1-4.13.

4. Use property 3, p. 11 to find the pseudoinverse of

$$A = \begin{bmatrix} -3 & 3 & 1 \\ 1 & -1 & 1 \\ 5 & -5 & 1 \end{bmatrix}.$$

5. Show that for any $m \times n$ matrix A and any $m \times 1$ vector b, either

$$Ax = b, \quad x \geq 0 \qquad \text{(a)}$$

has a solution x, or

$$A^T y \geq 0, \quad b^T y < 0 \qquad \text{(b)}$$

has a solution, but not both.

6. Show that for any $m \times n$ matrix A and any $m \times 1$ vector b, either

$$Ax \leq b, \quad x \geq 0 \qquad \text{(c)}$$

has a solution x, or

$$A^T y \geq 0, \quad b^T y < 0, \quad y \geq 0 \qquad \text{(d)}$$

has a solution y, but not both.

7. Show that for any $m \times n$ matrix A and any $m \times 1$ vector b, either

$$Ax \leq b \qquad \text{(e)}$$

has a solution x, or

$$A^T y = 0, \quad b^T y < 0, \quad y \geq 0 \qquad \text{(f)}$$

has a solution y, but not both.

For problem 8–14 consider the following definitions:

> *Definition:* A square matrix A is said to be an *EP* matrix provided $N(A) = N(A^T)$.
>
> *Definition:* An $n \times n$ matrix A (not necessarily symmetric) is positive semidefinite if and only if $x^T A x \geq 0$, for all $x \in E_n$.

8. Show that if A is positive semidefinite, then A is *EP*.

9. Show that if A is positive semidefinite and $BAB = B$, then B is positive semidefinite if and only if B is an *EP* matrix.

10. Use problem 9 to show that if A is positive semidefinite then so is A^+.

11. Let A be an $n \times n$ matrix and let B be such that $ABA = A$ and $BAB = B$; then show that $A^T B^T = AB$ if and only if A and B are *EP* matrices.

12. Use problem 11 to show that if A is positive semidefinite and B satisfies $BAB = B$, $ABA = A$, then B is positive semidefinite if and only if $A^T B^T = AB$.

13. Let A be an *EP* matrix and let U be any subspace of E_n complementary to $R(A)$. Show that there is a unique *EP* matrix B, which is a reflexive inverse of A having $U = N(B)$.

14. Suppose A is a positive semidefinite $n \times n$ matrix, C is an $r \times n$ matrix, and $\bar{A} = A + C^T C$. Let B be a positive semidefinite reflexive inverse of A and define

$$\bar{B} = B - BC^T(I + CBC^T)^{-1}CB.$$

Then show that $\bar{B}$ is a pseudoinverse of $\bar{A}$ if and only if $N(A) \subseteq N(C)$. If this is the case, then show

  a. $\overline{BA} = BA$ and $\overline{AB} = AB$
  b. $\bar{B}$ is positive semidefinite
  c. $\bar{B} = \bar{A}^+$ if and only if $B = A^+$.

CHAPTER 2

Statistical Concepts

2-1 Introductory Comments

The spirit of this book is nonparametric; that is, we attempt to eliminate the assumption that the random errors committed by taking observations must come from a population whose probability density function is known by the observer. However, in order to relate to the results obtained based on this latter assumption, we introduce a notation and some basic statistical concepts.

The concepts of *random vectors* and their *multivariate probability density functions* are basic to our discussion along with their *moments*. The idea of a criterion of excellence to evaluate the worth of a *statistic* used for *estimating* is introduced. It is assumed that the reader has at least a basic understanding of the univariate analogs; hence what we introduce here is admittedly minimal, but sufficient for our purposes.

2-2 Random Variables and Density Functions

If one can visualize a set of all possible measurements one could make during an experiment, it is possible to develop briefly the concept of a random variable. It is convenient to think of each outcome as a set of numerical measurements that has a random property. However, this is not usually the case; therefore, one would like to introduce a function that maps these outcomes to the real numbers. This function possessing special properties will be called a random variable, which we shall now formally formulate.

Definition 2-2.1. *A sample space* S *is the set of all possible outcomes of an experiment.*

An arbitrary individual outcome *s* will be referred to as an "elementary event." The set *S* is called the "sure event." Associated with the sure event *S* is a nonempty set *R* of subsets of *S* called a sigma algebra of the subsets of *S* defined as follows. $R: \sigma\text{-algebra}$

Definition 2-2.2. *If* R *is a set of subsets of* S *such that*

a. *If* $A_1, A_2, \ldots$ *are elements of* R, *then* $\cup_{i=1}^{\infty} A_i$ *is an element of* R.
b. *For every* A *an element of* R; *then the complement of* A, *denoted by* A^c, *is an element of* R, *and*
c. *0 is an element of* R; *then* R *is called a sigma algebra.*

The elements of *R* are called "events." With Definitions 2-2.1 and 2-2.2 in mind, we shall make the following definitions of a probability function and of random variables.

Definition 2-2.3. *A probability function* P *is a function that assigns to every event* A *a real number* P(A) *such that the following are true:*

a. $P(A) \geq 0$, *for every* $A \in R$,
b. $P(S) = 1$, *and*
c. *for every countable sequence of disjoint events* $A_1, A_2, \ldots, A_i$, *and so on,*

$$P(\cup_{i=1}^{\infty} A_i) = \sum_{i=1}^{\infty} P(A_i).$$

P(A) *is called the probability of* A.

Definition 2-2.4. *A random variable* X *is a function whose domain is the sample space* S *and whose range is a nonempty set of real numbers such that for any real number* a, *the set* $\{s \in S: X(s) \leq a\}$ *is an element of* R.

Definition 2-2.5. *A* $p \times 1$ *vector* $X = (X_1, X_2, \ldots, X_p)^T$ *is said to be a p-dimensional random vector or a random variable if* $X_1, X_2, \ldots, X_p$ *are random variables.*

For a thorough treatment of the concept of a random variable, see References [122] and [190].

Sec. 2-2 Random Variables and Density Functions

Definition 2-2.6. Let $X = (X_1, X_2, \ldots, X_p)^T$ be a p-dimensional random vector; then the joint distribution (multivariate distribution) of the random variables $X_1, X_2, \ldots, X_p$ is defined by

$$F_{X_1, X_2, \ldots, X_p}(x_1, x_2, \ldots, x_p) = P[\bigcap_{i=1}^{p} \{s \in S : X_i(s) \leq x_i\}]. \quad (2\text{-}1)$$

There are some interesting properties of the multivariate distribution, of which we list a few.

1. If $x_1 \leq x_1', x_2 \leq x_2', \ldots, x_p \leq x_p'$, then $F_{X_1, X_2, \ldots, X_p}(x_1, x_2, \ldots, x_p) \leq F_{X_1, X_2, \ldots, X_p}(x_1', x_2', \ldots, x_p')$.
2. $\lim_{x_i \to \infty} F_{X_1, X_2, \ldots, X_p}(x_1, x_2, \ldots, x_p)$
   $= F_{X_1, \ldots, X_{i-1}, X_{i+1}, \ldots, X_p}(x_1, \ldots, x_{i-1}, x_{i+1}, \ldots, x_p), \quad i = 1, 2, \ldots, p.$
3. $\lim_{x_i \to -\infty} F_{X_1, X_2, \ldots, X_p}(x_1, x_2, \ldots, x_p) = 0, \quad i = 1, 2, \ldots, p.$
4. $\lim_{\min x_i \to \infty} F_{X_1, X_2, \ldots, X_p}(x_1, x_2, \ldots, x_p) = 1.$

A random variable X is said to be discrete if its range is finite or denumerable; or more precisely, this means there exists a set of real numbers $X_1, X_2, \ldots, X_i$, and so on, finite or denumerable such that

$$\bigcup_i \{s \in S : X(s) = x_i\} \subset S.$$

The corresponding distribution function of the discrete random variable is defined as

$$F_X(x) = \sum P\{s \in S : X(s) = x_i\},$$

where the sum is taken over all i such that $x_i \leq x$. In this text, we are mainly interested in continuous random variables; therefore, the following definitions and theorems are of importance.

Definition 2-2.7. Let $X = (X_1, X_2, \ldots, X_p)^T$ be a p-dimensional random vector, then the random variables $X_1, X_2, \ldots, X_p$ are said to have a joint absolutely continuous distribution if there exists a function $f_{X_1, X_2, \ldots, X_p}(x_1, x_2, \ldots, x_p)$ such that

$$F_{X_1, X_2, \ldots, X_p}(x_1, \ldots, x_p)$$

$$= \int_{-\infty}^{x_1} \cdots \int_{-\infty}^{x_p} f_{x_1, \ldots, x_p}(y_1, \ldots, y_p) \, dy_1 \ldots dy_p \quad (2\text{-}2)$$

for every p-tuple of real numbers $(x_1, \ldots, x_p)$.

The function $f_{X_1, \ldots, X_p}(x_1, \ldots, x_p)$ is called the "joint density function" of the random variables $X_1, \ldots, X_p$.

Definition 2-2.8. Let $X = (X_1, \ldots, X_p)^T$ be a p-*dimensional random vector; the marginal density function of* $X_i, X_j, \ldots, X_q$ *is defined to be*

$$f_{X_i, X_j, \ldots, X_q}(x_i, x_j, \ldots, x_q)$$

$$= \int_{-\infty}^{\infty} \cdots \int_{-\infty}^{\infty} f_{X_1, X_2, \ldots, X_p}(x_1, \ldots, x_p) \, dx_1 \ldots dx_{i-1}$$

$$dx_{i+1} \ldots dx_{j-1} dx_{j+1} \ldots dx_{q-1} dx_{q+1} \ldots dx_p, \qquad (2\text{-}3)$$

where $1 \leq i, j, \ldots, q \leq p$.

For brevity we denote $X^{(2)} = (X_i, X_j, \ldots, X_q)^T$ and $X^{(1)} = (X_1, \ldots, X_{i-1}, X_{i+1}, \ldots, X_{j-1}, X_{j+1}, \ldots, X_{q-1}, X_{q+1}, \ldots, X_p)^T$; therefore, the left side of (2-3) can be denoted by

$$f_{X^{(2)}}(x^{(2)}).$$

Definition 2-2.9. Let $X = (X_1, \ldots, X_q)^T$ be a p-*dimensional random vector. Suppose we partition* X *as in Definition 2-2.8; then the conditional distribution of a random vector* $X^{(1)}$, *given that the random vector* $X^{(2)} = x^{(2)}$, *is defined by*

$$F_{X^{(1)}|X^{(2)}}(x^{(1)}|x^{(2)}) = \lim_{(\epsilon_i, \epsilon_j, \ldots, \epsilon_q \to 0)} P[\bigcap_{t \neq i, j, \ldots, q} \{s \in S : X_t(s) \leq x_t\} | \\ \bigcap_{z=1, j, \ldots, q} \{s \in S : x_z - \epsilon_z \leq X_z(s) \leq x_z + \epsilon_z\}], \qquad (2\text{-}4)$$

provided that such a limit exists. If

$$F_{X^{(1)}|X^{(2)}}(x^{(1)}|x^{(2)})$$

exists, then a conditional density $X^{(1)}$ *given* $X^{(2)}$ *denoted by*

$$f_{X^{(1)}|X^{(2)}}(x^{(1)}|x^{(2)}),$$

is given by the integral

$$F_{X^{(1)}|X^{(2)}}(x^{(1)}|x^{(2)})$$

$$= \int_{-\infty}^{x_1} \cdots \int_{-\infty}^{x_p} f_{X^{(1)}|X^{(2)}}(y^{(1)}|x^{(2)}) \, dy_1 \ldots dy_{i-1}, dy_{i+1} \ldots \\ dy_{j-1}, dy_{j+1} \ldots dy_{q-1}, dy_{q+1} \ldots dy_p. \qquad (2\text{-}5)$$

THEOREM 2-2.1. Let $X = (X_1, \ldots, X_p)$ be a p-*dimensional random vector. Suppose* X *has been partitioned as in Definition 2-2.8 and that* $X^{(1)}$

and $X^{(2)}$ have a joint absolutely continuous distribution; then at every point $(x^{(1)}, x^{(2)})$ at which $f_{X^{(1)},X^{(2)}}(x^{(1)}, x^{(2)})$ is continuous and $f_{X^{(2)}}(x^{(2)}) > 0$ and is continuous, there exists a conditional density of $X^{(2)}$ given $X^{(1)}$ such that

$$f_{X^{(1)}|X^{(2)}}(x^{(1)} | x^{(2)}) = \frac{f_{X^{(1)},X^{(2)}}(x^{(1)}, x^{(2)})}{f_{X^{(2)}}(x^{(2)})}. \tag{2-6}$$

Definition 2-2.10. *Two random vectors $X^{(1)}$ and $X^{(2)}$ are said to be statistically independent if and only if*

$$\begin{aligned} f_{X^{(1)}|X^{(2)}}(x^{(1)} | x^{(2)}) &= f_{X^{(1)}}(x^{(1)}), \\ f_{X^{(2)}|X^{(1)}}(x^{(2)} | x^{(1)}) &= f_{X^{(2)}}(x^{(2)}), \end{aligned} \tag{2-7}$$

where $f_{X^{(1)}}(x^{(1)})$ and $f_{X^{(2)}}(x^{(2)})$ are the marginal densities of $X^{(1)}$ and $X^{(2)}$, respectively.

THEOREM 2-2.2. *Continuous random vectors are statistically independent if and only if their joint density function is the product of their respective marginal density functions:*

$$f_{X^{(1)},X^{(2)}}(x^{(1)}, x^{(2)}) = f_{X^{(1)}}(x^{(1)}) f_{X^{(2)}}(x^{(2)}). \tag{2-8}$$

A very important and popular density function is the normal density function. We say that the real random vector $X = (X_1, X_2, \ldots, X_p)^T$ is distributed according to a normal density function if

$$f_X(x) = (2\pi)^{-p/2} |V|^{-1/2} \exp -\tfrac{1}{2}(x - \mu)^T V^{-1}(x - \mu) \tag{2-9}$$

$-\infty < x_i < \infty$, $-\infty < \mu_i < \infty$ for all i. The matrix V is positive definite. Note that

$$Q = (x - \mu)^T V^{-1}(x - \mu) \tag{2-10}$$

is a quadratic form. Since V is positive definite, V^{-1} is positive definite. Therefore, we have $Q > 0$ if $x \neq \mu$.

$$\text{Let } x = \begin{pmatrix} x^{(1)} \\ x^{(2)} \end{pmatrix}, \quad V = \begin{pmatrix} V_{11} & V_{12} \\ V_{21} & V_{22} \end{pmatrix}, \quad \text{and} \quad \mu = \begin{pmatrix} \mu^{(1)} \\ \mu^{(2)} \end{pmatrix} \tag{2-11}$$

be compatible partitioning of the matrices x, V, and μ, respectively.

$$\text{Let } R = \begin{pmatrix} R_{11} & R_{12} \\ R_{21} & R_{22} \end{pmatrix} = \begin{pmatrix} V_{11} & V_{12} \\ V_{21} & V_{22} \end{pmatrix}^{-1} = V^{-1}.$$

Then by Theorem 1-4.9 it follows that

$$R_{11} = V_{11}^{-1} + V_{11}^{-1}V_{12}R_{22}V_{21}V_{11}^{-1}$$

$$R_{12} = -V_{11}^{-1}V_{12}R_{22}$$

$$R_{21} = -R_{22}V_{21}V_{11}^{-1}$$

$$R_{22} = (V_{22} - V_{21}V_{11}^{-1}V_{12})^{-1}.$$

Substituting these identities into (2-10) and expanding by direct multiplication the exponent of $f_X(x)$, which (we denote by Q) one can obtain as follows:

$$\begin{aligned} Q &= (x^{(1)} - \mu^{(1)})^T(V_{11}^{-1} + V_{11}^{-1}V_{12}R_{22}V_{21}V_{11}^{-1})(x^{(1)} - \mu^{(1)}) \\ &\quad - (x^{(1)} - \mu^{(1)})^T V_{11}^{-1}V_{12}R_{22}(x^{(2)} - \mu^{(2)}) \\ &\quad - (x^{(2)} - \mu^{(2)})^T R_{22}V_{21}V_{11}^{-1}(x^{(1)} - \mu^{(1)}) \\ &\quad + (x^{(2)} - \mu^{(2)})R_{22}(x^{(2)} - \mu^{(2)}) \\ &= (x^{(1)} - \mu^{(1)})^T V_{11}^{-1}(x^{(1)} - \mu^{(1)}) \\ &\quad \times \{x^{(2)} - [\mu^{(2)} + V_{21}V_{11}^{-1}(x^{(1)} - \mu^{(1)})]\}^T(V_{22} - V_{21}V_{11}^{-1}V_{12})^{-1} \\ &\quad \times \{x^{(2)} - [\mu^{(2)} + V_{21}V_{11}^{-1}(x^{(1)} - \mu^{(1)})]\}. \end{aligned}$$

Using Theorem 1-4.13, we can write

$$|V|^{-1} = |V_{11}|^{-1}|V_{22} - V_{21}V_{11}^{-1}V_{12}|^{-1},$$

which allows us to write (2-9) as

$$\begin{aligned} f_X(x) &= (2\pi)^{-q/2}|V_{11}|^{-1/2}\exp\{-\tfrac{1}{2}(x^{(1)} - \mu^{(1)})^T V_{11}^{-1}(x^{(1)} - \mu^{(1)})\} \\ &\quad \times (2\pi)^{-(p-q)/2}|V_{22} - V_{21}V_{11}^{-1}V_{12}|^{-1/2}\exp -\tfrac{1}{2}\{(x^{(2)} \\ &\quad - [\mu^{(2)} + V_{21}V_{11}^{-1}(x^{(1)} - \mu^{(1)})]^T[V_{22} - V_{21}V_{11}^{-1}V_{12}]^{-1} \\ &\quad \times (x^{(2)} - [\mu^{(2)} + V_{21}V_{11}^{-1}(x^{(1)} - \mu^{(1)})]\}. \end{aligned}$$

Similarly, the normal density function can be written as

$$\begin{aligned} f(x) &= (2\pi)^{-(p-q)/2}|V_{22}|^{-1/2}\exp\{-\tfrac{1}{2}(x^{(2)} - \mu^{(2)})^T V_{22}^{-1}(x^{(2)} - \mu^{(2)})\} \\ &\quad \times (2\pi)^{-q/2}|V_{11} - V_{12}V_{22}^{-1}V_{21}|^{-1/2}\exp -\tfrac{1}{2}\{(x^{(1)} \\ &\quad - [\mu^{(1)} + V_{12}V_{22}^{-1}(x^{(2)} - \mu^{(2)})])^T[V_{11} - V_{12}V_{22}^{-1}V_{21}]^{-1} \\ &\quad \times (x^{(1)} - [\mu^{(1)} + V_{12}V_{21}(x^{(2)} - \mu^{(2)})])\}. \end{aligned}$$

Sec. 2-3 Expectations and Moments

Applying Definition 2-2.9, one can obtain from these characterizations of $f(x)$ that the marginal densities of the vectors $X^{(1)}$ and $X^{(2)}$ are

$$f_1(x^{(1)}) = (2\pi)^{-q/2} |V_{11}|^{-1/2} \exp -\tfrac{1}{2}\{(x^{(1)} - \mu^{(1)})^T V_{11}^{-1}(x^{(1)} - \mu^{(1)})\}$$

$$f_2(x^{(2)}) = (2\pi)^{-(p-q)/2} |V_{22}|^{-1/2} \exp -\tfrac{1}{2}\{(x^{(2)} - \mu^{(2)})^T V_{22}^{-1}(x^{(2)} - \mu^{(2)})\}.$$

It is important to note that $f_1(x^{(1)})$ and $f_2(x^{(2)})$ are normal density functions with the parameters μ and V replaced by $\mu^{(1)}$ and V_{11}, and $\mu^{(2)}$ and V_{22}, respectively.

Applying (2-10), one can obtain the conditional density functions of $X^{(1)}$ given $X^{(2)} = x^{(2)}$ and $X^{(2)}$ given $X^{(1)} = x^{(1)}$. These are, respectively,

$$f_1(x^{(1)} | x^{(2)}) = (2\pi)^{-q/2} |V_{11} - V_{12} V_{22}^{-1} V_{21}|^{-1/2} \exp -\tfrac{1}{2}\{(x^{(1)}$$
$$- [\mu^{(1)} + V_{12} V_{22}^{-1}(x^{(2)} - \mu^{(2)})])^T (V_{11} - V_{12} V_{22}^{-1} V_{21})^{-1}$$
$$\times (x^{(1)} - [\mu^{(1)} + V_{12} V_{22}^{-1}(x^{(2)} - \mu^{(2)})])\}$$

$$f_2(x^{(2)} | x^{(1)}) = (2\pi)^{-(p-q)/2} |V_{22} - V_{21} V_{11}^{-1} V_{12}|^{-1/2} \exp -\tfrac{1}{2}\{(x^{(2)}$$
$$- [\mu^{(2)} + V_{21} V_{11}^{-1}(x^{(1)} - \mu^{(1)})])^T (V_{22} - V_{21} V_{11}^{-1} V_{12})^{-1}$$
$$\times (x^{(2)} - [\mu^{(2)} + V_{21} V_{11}^{-1}(x^{(1)} - \mu^{(1)})])\}.$$

Thus it follows that

$$f(x) = f_1(x^{(1)} | x^{(2)}) f_2(x^{(2)})$$
$$= f_2(x^{(2)} | x^{(1)}) f_1(x^{(1)}),$$

and that the conditional density function of $X^{(1)}$ given $X^{(2)} = x^{(2)}$ and $X^{(2)}$ given $X^{(1)}$ are normal density functions with the parameter μ replaced by the $\mu^{(1)} + V_{12} V_{22}^{-1}(x^{(2)} - \mu^{(2)})$ and $\mu^{(2)} + V_{21} V_{11}^{-1}(x^{(1)} - \mu^{(1)})$, respectively, and the parameter V replaced by $V_{11} - V_{12} V_{22}^{-1} V_{21}$ and $V_{22} - V_{21} V_{11}^{-1} V_{12}$, respectively.

2-3 Expectations and Moments

In this section we introduce the notion of the expected value. We shall define it so that it can be applied to any type of random variable, particularly discrete and continuous random variables.

Definition 2-3.1. *Let the domain of the real-valued function* g *be a p-dimensional space of real numbers. Then the expected value of the function* g *of the random variables* $X_1, X_2, \ldots, X_p$ *is defined to be*

$$E[g(X_1, X_2, \ldots, X_p)]$$
$$= \int_{x_1} \cdots \int_{x_p} g(x_1, x_2, \ldots, x_p) \, dF_{X_1, X_2, \ldots, X_p}(x_1, x_2, \ldots, x_p) \quad (2\text{-}12)$$

whenever the integral exists. (See Reference [87], [133], and [190].)

Since we are mainly interested in continuous random variables, the integral in (2-12) reduces to

$$E[g(X_1, X_2, \ldots, X_p)]$$
$$= \int_{x_1} \cdots \int_{x_p} g(x_1, \ldots, x_p) f_{X_1, \ldots, X_p}(x_1, x_2, \ldots, x_p) \, dx_1 \cdots dx_p \quad (2.13)$$

whenever the distribution function $F_{X_1, \ldots, X_p}(x_1, \ldots, x_p)$ is jointly absolutely continuous. If $g(X_1, \ldots, X_p) = X_i$, $1 \le i \le p$, then

$$E[g(X_1, \ldots, X_p)] = \int_{x_i} \cdots \int_{x_p} x_i \, dF_{X_1, \ldots, X_p}(x_1, \ldots, x_p).$$

We call $E[X_i]$ the first *moment* or *mean value* of the random variable X_i. If

$$g(X_1, \ldots, X_p) = X_i^k, \quad 1 \le i \le p,$$

then we call $E(X_i^k)$ the *kth moment* of the random variable X_i.

With Definition 2-3.1 in mind, we now define what we mean by the expected value of a $(p \times 1)$ random vector $X = (X_1, \ldots, X_p)^T$.

Definition 2-3.2. *The expected value of a* (p × 1) *random vector* $X = (X_1, \ldots, X_p)^T$ *is*

$$E(X) = [E(X_1), E(X_2), \ldots, E(X_p)]^T. \quad (2\text{-}14)$$

The vector (2-14) *is called the* **mean** *vector of the random vector* X.

Another useful definition is as follows.

Definition 2-3.3. *The covariance matrix of a random vector* $X = (X_1, \ldots, X_p)^T$ *is the positive definite matrix*

$$E\{[X - E(X)][X - E(X)]^T\} \quad (2\text{-}15)$$

whenever it exists. The ith diagonal element yields the variance of X_i, *and the ijth component yields the covariance of* X_i *and* X_j.

THEOREM 2-3.1. Let X be a (p × 1) *random variable distributed according to the normal density function; then* E(X) *is the parameter* μ *and the covariance matrix is the parameter* V = (σ_{ij}).

Proof: Consider

$$E(X_i) = k \int_{-\infty}^{\infty} \cdots \int_{-\infty}^{\infty} x_i e^{-1/2(x-\mu)^T V^{-1}(x-\mu)} \, dx_1 \ldots dx_p$$

$$= \int_{-\infty}^{\infty} x_i \left[\int_{-\infty}^{\infty} k x_i e^{-1/2(x-\mu)^T V^{-1}(x-\mu)} \, dx_1 \ldots dx_{i-1} dx_{i+1} \ldots dx_p \right] dx_i$$

$$= \int_{-\infty}^{\infty} x_i (2\pi \sigma_{ii})^{-1/2} e^{-1/2\sigma_{ii}(x_i - \mu_i)^2} \, dx_i$$

$$= \mu_i,$$

where $k = (2\pi)^{-p/2} |V|^{-1/2}$. That is, $E(X) = \mu$, where $\mu = [\mu_1, \ldots, \mu_p]^T$.
Consider

$$E\{[X - E(X)][X - E(X)]^T\} = E[(X - \mu)(X - \mu)^T]$$

$$= \{E(X_i - \mu_i)(X_j - \mu_j)\}.$$

The elements of the covariance matrix are by definition the covariance of X_i and X_j. Clearly, if $i = j$, $E(X_i - \mu_i)^2$ is by definition the variance of X_i. That is, the ith diagonal element of the covariance matrix is the variance of X_i and the ijth element is the covariance of X_i and X_j ($i \neq j$).

Let Y be a $p \times 1$ random vector distributed according to a normal density function with parameters $\mu = 0$ and $V = I$, the $p \times p$ identity matrix. This statement we denote symbolically by

$$Y \sim N(0, I).$$

That is,

$$f(y) = (2\pi)^{-p/2} e^{-1/2 y^T y}, \quad -\infty < y_i < \infty, \quad i = 1, \ldots, p$$

$$= 0,$$

elsewhere. By direct integration, one can establish

$$E(Y_i) = 0,$$

$$E(Y_i Y_j) = 0, \quad \text{for } i \neq j,$$

$$E(Y_i^2) = 1, \quad i, j = 1, 2, \ldots, p.$$

We note that the variance is simply $E(Y_i^2) = 1$.

Suppose that we define

$$Y = P(X - \mu), \qquad (2\text{-}16)$$

where P is a nonsingular matrix such that

$$V^{-1} = P^T P \quad \text{or} \quad V = P^{-1} P^{-T}. \qquad (2\text{-}17)$$

Such a matrix always exists since V is positive definite, and in turn V^{-1} is positive definite (see Theorem 1-4.1).

Then the density function of Y is

$$g(y) = f(x) \left\| \frac{dy}{dx} \right\|^{-1}, \qquad (2\text{-}18)$$

where

$$\frac{dy}{dx} = \left\{ \frac{\partial y_i}{\partial x_j} \right\}$$

and $\|\cdot\|$ denotes the absolute value of the determinant of the $p \times p$ matrix. Since

$$Y = P(X - \mu)$$

and

$$\frac{dY}{dX} = P$$

it follows from (2-18) that if $X \sim N(\mu, V)$, then

$$g(y) = (2\pi)^{-p/2} |P^T P|^{+1/2} e^{-1/2 \cdot y^T y} |P|^{-1}.$$

But

$$|P^T P|^{+1/2} = |P^T|^{+1/2} |P|^{+1/2} = |P|^{+1},$$

which implies

$$g(y) = (2\pi)^{-p/2} e^{-1/2 \cdot y^T y}.$$

Since $E(Y) = PE(X - \mu) = 0$, and P is nonsingular, it follows that

$$E(X - \mu) = 0$$

or

$$E(X) = \mu.$$

Also,
$$E(YY^T) = I$$
or
$$E[P(X - \mu)(X - \mu)^T P^T] = I,$$
which implies
$$PE\{(X - \mu)(X - \mu)^T\}P^T = I$$
or
$$E\{(X - \mu)(X - \mu)^T\} = P^{-1}P^{-T} = (P^T P)^{-1}.$$

Definition 2-3.4. *The expectation*

$$C_X(t) = E(e^{it^T X}) \qquad (2\text{-}19)$$

of the (p × 1) *random vector X for all real vectors* $t = (t_1, t_2, \ldots, t_p)^T$ *is called the characteristic function of X.*

As an illustration of the characteristic function consider the following two examples. First, let the $(p \times 1)$ random vector Y be distributed $N(0, I)$; then

$$\begin{aligned}
C_Y(t) &= (2\pi)^{-p/2} \int_{-\infty}^{\infty} e^{it^T y} e^{-1/2 \cdot y^T y} \, dy \\
&= \prod_{j=1}^{P} \int_{-\infty}^{\infty} (2\pi)^{-1/2} e^{-1/2(y_j^2 - 2it_j y_j)} \, dy_j \\
&= \prod_{j=1}^{P} e^{-t_j^2/2}
\end{aligned} \qquad (2\text{-}20)$$

$$C_Y(t) = e^{-t^T t/2}.$$

Secondly, let $X \sim N(\mu, V)$. Then, letting $Y = P(X - \mu)$, it follows that $X = P^{-1} Y + \mu$. Hence

$$\begin{aligned}
C_X(s) &= E[e^{is^T X}] \\
&= E[e^{is^T P^{-1} Y + is^T \mu}] \\
&= e^{+is^T \mu} E[e^{-is^T P^{-1} Y}] \\
&= e^{is^T \mu} E[e^{-i(P^T s)^T Y}].
\end{aligned}$$

Letting $t = P^{-T}s$, it follows from (2-20) that

$$C_X(s) = e^{is^T\mu} e^{-s^T P^{-1} P^{-T} s/2}$$

$$= e^{is^T\mu - s^T V s/2}. \qquad (2\text{-}21)$$

An important relationship that makes the characteristic function indeed useful is summarized in Theorem 2-3.2.

THEOREM 2-3.2. *There exists a one-to-one correspondence between the set of all distribution functions and the set of all characteristic functions.*

It is clear from the definition that, if one knows the density function, he can obtain the characteristic function simply by performing a multiple integration. Theorem 2-3.3 tells one how to find the density function if he knows the characteristic function.

THEOREM 2-3.3. *If the random vector X has the density* $f(x)$ *and the characteristic function* $C_x(t)$, *then*

$$f(x) = \frac{1}{(2\pi)^p} \int_{-\infty}^{\infty} \cdots \int_{-\infty}^{\infty} e^{-it^T x} C_X(t)\, dt_1 \ldots dt_p. \qquad (2\text{-}22)$$

Theorem 2-3.3 and the definition of the characteristic function tell us that the density function and characteristic function are Fourier transform pairs.

Theorem 2-3.4 is useful.

THEOREM 2-3.4. *Let X be a random vector whose mean is* μ *and whose covariance matrix is* V_x. *Let A and B be matrices whose dimensions are such that*

$$Y = AX + B$$

is well defined; then

$$EY = A\mu + B,$$

and the covariance matrix V_Y *of Y is given by*

$$V_Y = A V_X A^T.$$

Proof: Since $E(\cdot)$ is a linear operator, it follows directly that

$$EY = A\mu + B.$$

Consider

$$V_Y = E[(Y - EY)(Y - EY)^T]$$
$$= E[AX + B - A\mu - B)(AX + B - A\mu - B)^T]$$
$$= E[A(X - \mu)(X - \mu)^T A^T]$$
$$= AE[(X - \mu)(X - \mu)^T]A^T$$
$$= AV_X A^T,$$

the desired result.

COROLLARY 2-3.1. *If* $X \sim N(\mu, V_x)$, *then* $Y = AX + B \sim N(A\mu + B, AV_x A^T)$, *where rank* A *(an* n $\times$ p *matrix) is* n $\leq$ p.

Proof: Since $X \sim N(\mu, V_x)$, it follows from the definition of $C_y(t)$, that

$$C_y(t) = E[e^{it^T(AX+B)}]$$
$$= e^{it^T B} E[e^{it^T AX}] \qquad (2\text{-}23)$$
$$= e^{it^T B} E[e^{is^T X}],$$

where $s = A^T t$. But by (2-22),

$$E(e^{is^T X}) = e^{is^T \mu - s^T V s/2}$$
$$= e^{it^T A\mu - t^T AVA^T t/2}.$$

Substituting this result into the left-hand side of (2-23) one obtains

$$C_y(t) = e^{it^T B + it^T A\mu - t^T AVA^T t/2}$$
$$= e^{it^T(A\mu+B) - t^T(AVA^T)t/2}.$$

Thus $Y \sim N(A\mu + B, AVA^T)$. We note that if rank of A is $r < \min[n, p]$, AVA^T is positive semidefinite. This case is known as singular normal and fails to fit the definition used here.

2-4 Estimators

Let θ be a parameter (a state vector unknown to the observer that one wishes to *estimate* using observations or data usually corrupted by *random error*, sometimes called noise. To illustrate let Y be a random variable such that $Y = \theta + e$, where θ is an unknown parameter and e is a random variable with zero mean and known variance. Suppose a random sample $(y_1, y_2, \ldots, y_n)$ of size n has been observed. Then it seems natural to use an estimate of the unknown parameter θ, which is the mean of Y, the sample mean $\bar{y}$ of the sample values; that is,

$$\bar{y} = \frac{1}{n} \sum_{i=1}^{n} y_i.$$

We emphasize that this sample mean $\bar{y}$ is called an *estimate* of the parameter θ. On the other hand before the random sample of size is taken, the sample values can be thought of as being random variables. The sample mean is now a random variable and is written as

$$\bar{Y} = \frac{1}{n} \sum_{i=1}^{n} Y_i.$$

We call the sample mean $\bar{Y}$ an *estimator* of the parameter. With this illustration in mind we give the following definition.

Definition 2-4.1. *An estimator $\hat{\Theta}$, sometimes called a statistic, is a function of the observation vector $Y = (Y_1, Y_2, \ldots, Y_n)^T$ not dependent on the parameter θ, which estimates θ.*

We use the word *estimates* in a similar fashion as the numerical analyst uses the word *approximates*.

Definition 2-4.2. *An estimator $\hat{\Theta}$ is said to be unbiased if and only if*

$$E[\hat{\Theta}] = \theta. \qquad (2\text{-}24)$$

Definition 2-4.3. *An estimator $\hat{\Theta}$ is a linear estimator of θ if*

$$\hat{\Theta} = BY, \qquad (2\text{-}25)$$

where B *is a matrix.*

It is important to note that one can select the matrix B so that the estimator is "best" according to a specified criteria of excellence. Heuristically, it is desirable to select an estimator $\hat{\Theta}$ so that the corresponding estimates are close to the unknown parameter θ. A better statement would be that the estimator $\hat{\Theta}$ should be selected so that for any other estimator Θ^* the probability

$$P[|\hat{\Theta} - \theta| < \epsilon] \geq P[|\Theta^* - \theta| < \epsilon] \qquad (2\text{-}26)$$

is true for every $\epsilon > 0$. This criterion is difficult to use in searching for an estimator $\hat{\Theta}$ to be best according to (2-26). A similar criterion called the *minimum variance criterion* is easier to work with and contains many of the same desirable properties as (2-26).

Definition 2-4.4. *The estimator $\hat{\Theta}$ is said to be best if $E(\hat{\Theta}) = \theta$ and for any other unbiased estimator of θ, say, Θ^*,*

$$E[(\hat{\Theta} - \theta)^2] \leq E[(\Theta^* - \theta)^2]. \qquad (2\text{-}27)$$

$\hat{\Theta}$ *is called a minimum variance unbiased estimator of θ.*

Note that one must insist that $\hat{\Theta}$ be unbiased for if one selects $\hat{\Theta}$ to be simply minimum variance, then $\hat{\Theta} = C$, where C is a constant, implies that the variance of $\hat{\Theta}$ is zero. We shall denote

$$E[(\hat{\Theta} - \theta)^2] = V_{\hat{\Theta}},$$

and note that

$$V_{\hat{\Theta}} = E(\hat{\Theta}^2) - [E(\hat{\Theta})]^2.$$

Note also that if

$$E(\hat{\Theta}) = \theta_1 \neq \theta,$$

then

$$E[(\hat{\Theta} - \theta)^2] = E[(\hat{\Theta} - \theta_1 + \theta_1 - \theta)^2]$$
$$= E[(\hat{\Theta} - \theta_1)^2 + (\theta_1 - \theta)^2]$$
$$= V_{\hat{\Theta}} + (\text{bias})^2,$$

where

$$\text{bias} = (\theta_1 - \theta).$$

Whenever the requirement of unbiased estimators is removed, then $E[(\hat{\Theta} - \theta)^2]$ is called the *mean-squared error*. If the restriction of requiring $\hat{\Theta}$ to be an unbiased estimator of θ is removed, then estimators of θ that minimize the mean-squared error are sometimes desirable. However, for most densities $f(X; \theta)$, there does not exist an estimator that minimizes the mean-squared error for all values of θ, but an estimator may exist that minimizes the mean-squared error for one set of values of θ and another estimator may exist that minimizes the mean-squared error for another set of values of θ. This makes the problem of finding an estimator of θ, which minimizes the mean-squared error, considerably more difficult since θ is unknown.

Definition 2-4.5. *A covariance matrix* V_1 *is said to be smaller than the covariance matrix* V_2 *if* $V = V_2 - V_1$ *is positive semidefinite.*

It is important to note that the diagonal elements of V must be non-zero, or if $v_{ii} = 0$, then $v_{ji} = v_{ij} = 0$ for all j.

Next it seems desirable that, as the random sample is made larger, then the corresponding estimator $\hat{\Theta}$ becomes better in some sense. To be more specific suppose we have a set of estimators $\hat{\Theta}_1, \hat{\Theta}_2, \ldots, \hat{\Theta}_n$, and so on, based on corresponding random samples of sizes $1, 2, \ldots n$, and so on; then in probability one desires that

$$P[|\hat{\Theta}_n - \theta| < \epsilon] \to 1 \quad \text{as } n \to \infty \tag{2-28}$$

for every $\epsilon > 0$. The symbol $\hat{\Theta}_n$ denotes that $\hat{\Theta}$ is a function of the sample size n. Whenever we have a set of estimators $\hat{\Theta}_1, \hat{\Theta}_2, \ldots, \hat{\Theta}_n$, and so on, such that (2-28) holds, then this sequence is called a *simple consistent estimator of* θ.

Finally, two popular criteria for selecting estimators are the *maximum likelihood* criterion and the so-called *Bayesian criterion* which minimize a preassigned loss function (see Definition 2-4.6). To obtain these estimators, one must know the exact form of various density functions; hence they are parametric in nature.

Definition 2-4.6. *Let* $L = L(y_1, y_2, \ldots, y_n; \theta)$ *be the joint probability density function of the observations depending on the unknown parameters* $\theta = (\theta_1, \theta_2, \ldots, \theta_p)$; *then the functions of* $y_1, \ldots, y_n$ [*say* $\hat{\Theta}_1 = \theta_2(y_1, \ldots, y_n), \hat{\Theta}_2 = \theta_2(y_1, \ldots, y_n), \ldots, \hat{\Theta}_p = \theta_p(y_1, \ldots, y_n)$), *such that* $L(y_1, \ldots, y_n; \theta)$ *is maximized whenever* $\theta = (\theta_1, \theta_2, \ldots, \theta_p)$ *is replaced by* $\hat{\Theta} = (\hat{\Theta}_1, \hat{\Theta}_2, \ldots, \hat{\Theta}_p)$] *are called maximum likelihood estimators for* θ_i, $i = 1, 2, \ldots, p$.

In many instances there is an unique set $\hat{\Theta} = (\hat{\Theta}_1, \hat{\Theta}_2, \ldots, \hat{\Theta}_p)$ that maximizes the likelihood function L, and often this set can be obtained by a process of differentiation. As an illustration consider

$$X^{(j)} \sim N(\mu, V), \quad j = 1, 2, \ldots, N,$$

where $X^{(j)}$ denotes a $p \times 1$ random vector and the set $\{X^{(j)}, j = 1, 2, \ldots, n\}$ denotes an *independent random* sample. The likelihood function is

$$L(x^{(1)}, \ldots, x^{(n)}) = (2\pi)^{-np/2} |V|^{-n/2} \exp\left[-\tfrac{1}{2} \sum_{j=1}^{n} (x^j - \mu)^T V^{-1} (x^j - \mu)\right].$$

Our desire is to select two estimators, say, $\hat{\mu}$ and $\hat{V}$, which maximize the likelihood function. Theorem 2-4.1 leads to the desired result.

Sec. 2-4   Estimators

THEOREM 2-4.1. *The values of μ and V, which maximize $\ln L$, also maximize L.*

The proof of this theorem follows from the fact that $\ln L$ is a monotone increasing function of x.

Consider then

$$\ln L = -\frac{np}{2} \ln (2\pi) - \frac{n}{2} \ln |V|$$

$$- \frac{1}{2} \sum_{j=1}^{n} (x^j - \mu)^T V^{-1} (x^j - \mu)$$

$$= -\frac{np}{2} \ln (2\pi) - \frac{n}{2} \ln |V|$$

$$- \frac{1}{2} \sum_{j=1}^{n} (x^j - \bar{x})^T V^{-1} (x^j - \bar{x})$$

$$- \frac{n}{2} (\bar{x} - \mu)^T V^{-1} (\bar{x} - \mu).$$

Note that $\ln L$ is maximized if the positive definite quadratic form

$$(\bar{x} - \mu)^T V^{-1} (\bar{x} - \mu) = 0.$$

This can be true only if $(\bar{x} - \mu) = 0$ or $\bar{x} = \mu$, that is, if $\hat{\mu} = \bar{x}$ is the maximum likelihood estimate for μ. Thus the maximum likelihood estimator of μ is $\hat{\mu} = \bar{X}$. To find $\hat{V}$ consider Theorems 2-4.2 and 2-4.3.

THEOREM 2-4.2. *Let*

$$f(C) = \tfrac{1}{2} N \log |C| - \tfrac{1}{2} \operatorname{tr} CD,$$

where $C = \{c_{ij}\}$ is a positive semidefinite matrix, $D = \{d_{ij}\}$ is a positive definite matrix, and N is a scalar. Then the maximum of $f(C)$ is taken on at $C = ND^{-1}$.

Proof: A necessary condition for f to possess a critical value is that (see Theorem 1-6.2)

$$\frac{\partial f}{\partial c_{kk}} = \frac{1}{2} \frac{N}{|C|} \frac{\partial |C|}{\partial c_{kk}} - \frac{1}{2} d_{kk}$$

$$= \frac{1}{2} N \frac{\operatorname{cof} c_{kk}}{|C|} - \frac{1}{2} d_{kk} = 0 \qquad (2\text{-}29)$$

$$k = 1, 2, \ldots, n$$

and when $k \neq e$, then

$$\frac{\partial f}{\partial c_{ke}} = \frac{N}{|C|} \operatorname{cof} c_{ke} - \frac{1}{2}(2d_{ke})$$

$$= N \frac{\operatorname{cof} c_{ke}}{|C|} - d_{ke} = 0 \qquad (2\text{-}30)$$

$$k = 1, 2, \ldots, n$$

$$e = 1, 2, \ldots, n$$

$$k \neq e,$$

since $c_{ke} = c_{ek}$. In matrix notation the n^2 equations defined by (2-29) and (2-30) are

$$NC^{-1} = D,$$

since

$$C^{-1} = \frac{\{\operatorname{cof} c_{ke}\}^T}{|C|},$$

which implies that

$$C^{-1} = \frac{D}{N}$$

or

$$C = ND^{-1},$$

the desired result.

Note that

$$\ln L = -\frac{np}{2} \ln(2\pi) - \frac{n}{2} \ln|V| - \frac{1}{2} \sum_{j=1}^{n} (x^j - \bar{x})^T V^{-1}(x^j - \bar{x}) \qquad (2\text{-}31)$$

can be written as

$$\ln L = -\frac{np}{2} \ln(2\pi) + \frac{n}{2} \ln|V^{-1}| - \operatorname{tr} V^{-1} A$$

$$- \frac{1}{2} n(\bar{x} - \hat{\mu})^T V^{-1}(\bar{x} - \hat{\mu}).$$

This result follows directly from (2-31) if one applies Theorem 2-4.3.

THEOREM 2-4-3.

$$\sum_{n}^{j} (x^j - \mu)^T V^{-1}(x^j - \mu) = \text{tr } V^{-1}A + n(\bar{x} - \mu)^T V^{-1}(\bar{x} - \mu).$$

Proof: Using the fact that tr $CD = $ tr DC, and that the trace of a scalar is the scalar itself, one can proceed:

$$\sum_{j=1}^{n} (x^j - \mu)^T V^{-1}(x^j - \mu)$$

$$= \text{tr} \sum_{j=1}^{n} (x^j - \mu)^T V^{-1}(x^j - \mu)$$

$$= \text{tr} \sum_{j=1}^{n} V^{-1}(x^j - \mu)(x^j - \mu)^T$$

$$= \text{tr} \sum_{j=1}^{n} V^{-1}[(x^j - \bar{x}) + (\bar{x} - \mu)][(x^j - \bar{x}) + (\bar{x} - \mu)]^T$$

$$= \text{tr} \sum_{j=1}^{n} [V^{-1}(x^j - \bar{x})(x^j - \bar{x})^T + V^{-1}(\bar{x} - \mu)(\bar{x} - \mu)^T]$$

$$= \text{tr } V^{-1} \sum_{j=1}^{n} (x^j - \bar{x})(x^j - \bar{x})^T + \text{tr } nV^{-1}(\bar{x} - \mu)(\bar{x} - \mu)^T$$

$$= \text{tr } V^{-1}A + n(\bar{x} - \mu)^T V^{-1}(\bar{x} - \mu),$$

where

$$A = \sum_{j=1}^{n} (x^j - \bar{x})(x^j - \bar{x})^T.$$

But if $\hat{\mu} = \bar{x}$, then $(\bar{x} - \hat{\mu})^T V^{-1}(\bar{x} - \hat{\mu}) = 0$. Hence our problem reduces to maximizing

$$f(V^{-1}) = -\frac{np}{2} \ln (2\pi) + \frac{n}{2} \ln |V^{-1}| - \text{tr } V^{-1} A.$$

If we let $C = V^{-1}$, then Theorems 2-4.1. and 2-4.2 imply that $f(V^{-1})$ is maximized if

$$\hat{\mu} = \bar{x}$$

and

$$\hat{V}^{-1} = \left(\frac{1}{n} A\right)^{-1}.$$

It is true, but perhaps not obvious, that this implies that

$$\hat{V} = \frac{A}{n},$$

or

$$\hat{V} = \sum_{j=1}^{n} \frac{(x^j - \bar{x})(x^j - \bar{x})^T}{n}.$$

Thus the maximum likelihood estimator of V is

$$\hat{V} = \frac{1}{n} \sum_{j=1}^{n} (X^j - \bar{X})(X^j - \bar{X})^T.$$

The discriminating feature in determining a Bayesian estimator for a parameter Θ, is that one assumes the parameter Θ is a random variable with a known probability density function. It is important to note that the requirement that one must model the population of Θ's by defining the form of Θ's density function with known moments is the major disadvantage of this technique.

Let the joint probability density function of the observations $Y = (Y_1, \ldots, Y_n)^T$ and the parameter, Θ, be

$$f(y, \theta) = h_{Y|\Theta}(y|\theta) g_\Theta(\theta)$$

$$= h_{\Theta|Y}(\theta|y) g_Y(y),$$

where h_Y and h_Θ denote the conditional density function of Y given Θ and Θ given Y, respectively, while g_Θ and g_Y are the marginal density functions of Θ and Y, respectively.

Definition 2-4.7. *A loss function* $l(\hat{\Theta}, \theta)$ *is a nonnegative function such that*

1. $l(\hat{\Theta}, \theta) \geq 0$ *for all admissible values of* Θ *and* $\hat{\Theta}$.
2. *For each admissible value of* Θ *there is at least one* $\hat{\Theta}$ *such that* $l(\hat{\Theta}, \theta) = 0$.

Definition 2-4.8. *Let* $\Theta \sim g_\Theta(\theta)$ *and* $Y \sim h_Y(y|\theta)$; *then the Bayes estimator for* Θ, *say,* $\hat{\Theta}$, *is that statistic* $\hat{\Theta} = \hat{\Theta}(Y_1, Y_2, \ldots, Y_n)$ *such that the expected value of the loss function is minimized; that is, minimize*

$$E[l(\hat{\Theta}; \theta)] = \int_\theta \left\{ \int_{y_1} \cdots \int_{y_n} l(\hat{\Theta}; \theta) h_{Y|\Theta}(y|\theta) \, dy_1 \ldots dy_n \right\} g_\Theta(\theta) \, d\theta.$$

Note

$$h_{\Theta|Y}(\theta|y) = \frac{f(y, \theta)}{g_Y(y)} = \frac{h_{Y|\Theta}(y|\theta)g_\Theta(\theta)}{g_Y(y)},$$

which implies

$$E[l(\Theta: \theta)] = \int_{y_1} \cdots \int_{y_n} \left[\int_\theta l(\hat{\Theta}; \theta)h_{\Theta|Y}(\theta|y)\, d\theta\right] g_Y(y)\, dy_1 \cdots dy_n.$$

Now the problem of minimizing $E[l(\hat{\Theta}; \theta)]$ when $Y = y = (y_1, \ldots, y_n)$ is equivalent to minimizing

$$V(\hat{\Theta}, y) = \int_\theta l(\hat{\Theta}; \theta)h_{\Theta|Y}(\Theta|y)d\theta.$$

The function $V(\hat{\Theta}, y)$ is sometimes called the *a posteriori risk* for estimating Θ.

THEOREM 2-4.4. *Let* $l(\hat{\Theta}; \theta) = (\hat{\Theta} - \theta)^2$. *Then the estimate of* θ *that minimizes the expected loss function* $l(\hat{\Theta}; \theta)$ *is* $\hat{\Theta} = E(\Theta|Y)$, *the conditional expectation of* Θ *given* Y.

Proof: Consider

$$V(\hat{\Theta}, y) = \int_{-\infty}^{\infty} (\hat{\Theta} - \theta)^2 h_\Theta(\theta|y)\, d\theta.$$

Let $h_\Theta(\theta|y)$ be such that the operator $\partial/\partial\hat{\Theta}$ commutes with the integral operator; then

$$\frac{\partial v}{\partial \hat{\Theta}} = \int_{-\infty}^{\infty} 2(\hat{\Theta} - \theta)h_\theta(\Theta|y)\, d\theta.$$

A necessary condition for V to be minimal with respect to $\hat{\Theta}$ is that $\partial v/\partial \hat{\Theta} = 0$, which implies that

$$\hat{\Theta} = \int_{-\infty}^{\infty} \theta h_\Theta(\Theta|y)\, d\theta = E(\Theta|Y),$$

the desired results.

These concepts can be extended to matrix-valued loss functions, such as

$$l(\hat{\Theta}, \theta) = (\hat{\Theta} - \Theta)(\hat{\Theta} - \Theta)^T,$$

where $\hat{\Theta}$ and Θ are $p \times 1$ vectors, and $l(\hat{\Theta}, \theta)$ is a $p \times p$ matrix. Fortunately, in most cases

$$E(\hat{\Theta} - \Theta)(\hat{\Theta} - \Theta)^T$$

is a positive definite matrix, and as a function of some parameter, it can be minimized with respect to that parameter in the sense discussed in Section 1-7. For a more complete discussion of statistical concepts see References [8], [41], [87], [104], [106], and [133].

EXERCISES

1. Given the set $S = \{0, 2, 3, 5, 6\}$ form a sigma field of events.

2. Show that if $A_1, A_2, \ldots, A_n$, and so on, is a denumerable sequence of sets belonging to a sigma field of events, then

$$\bigcap_{n=1}^{\infty} A_n$$

is an element of the sigma field.

3. Let A_1, A_2, A_3, and A_4 be any four events; then show

$$P(A_1 \cup A_2 \cup A_3 \cup A_4) = P(A_1) + P(A_2) + P(A_3) + P(A_4)$$
$$- P(A_1 A_2) - P(A_1 A_3) - P(A_1 A_4) - P(A_2 A_3)$$
$$- P(A_2 A_4) - P(A_3 A_4) + P(A_1 A_2 A_3) + P(A_1 A_2 A_4)$$
$$+ P(A_1 A_3 A_4) + P(A_2 A_3 A_4) - P(A_1 A_2 A_3 A_4).$$

Write a general formula for $P\left(\bigcup_{i=1}^{n} A_i\right)$ whenever $A_1, A_2, \ldots, A_n$ are any n events.

4. Let (x_1, y_1) and (x_2, y_2) be a pair of values taken from a bivariate population with density

$$f(x, y) = \frac{1}{2\pi(1 - p^2)^{1/2}} \exp\left\{\frac{1}{2(1 - p^2)}(x^2 - 2pxy + y^2)\right\}$$

Define

$$\operatorname{sgn} t = \begin{cases} 1, & t > 0 \\ 0, & t = 0 \\ -1, & t < 0. \end{cases}$$

Assume the following result:

$$\text{sgn } t = \frac{1}{\pi} \int_{-\infty}^{\infty} \frac{e^{ipt}}{ip} \, dp,$$

and then find

$$E\{\text{sgn}(x_1 - x_2) \text{ sgn}(y_1 - y_2)\}.$$

5. Assuming the population in Exercise 4, let $(x_1, y_1), (x_2, y_2), \ldots, (x_n, y_n)$ be a random sample from this population. To each pair x_i, x_j is allotted a score $\text{sgn}(x_j - x_i) = a_{ij}$ and to the corresponding pair $y_i y_j$ is allotted the score $\text{sgn}(y_j - y_i) = b_{ij}$. Defining a rank correlation coefficient t as

$$t = \sum_{i,j=1}^{n} \frac{a_{ij} b_{ij}}{n(n-1)},$$

show that an estimator of p is given by $\sin \frac{1}{2} t$.

6. Pitman has proposed to define an estimator $\hat{\Theta}_1$ of a parameter as "closer" than an estimator $\hat{\Theta}_2$ if

$$\text{prob}(|\hat{\Theta}_1 - \theta| < |\hat{\Theta}_2 - \theta|) > \tfrac{1}{2}.$$

Show that if two unbiased estimators are jointly distributed in a normal form with variance $\sigma_{\hat{\Theta}_1}^2, \sigma_{\hat{\Theta}_2}^2$ and correlation p:

$$\text{prob}[|\hat{\Theta}_1 - \theta| < |\hat{\Theta}_2 - \theta|]$$

$$= \frac{1}{\pi} \arctan \frac{2\sigma_{\hat{\Theta}_1} \sigma_{\hat{\Theta}_2}(1 - p^2)^{1/2}}{\sigma_{\hat{\Theta}_1}^2 \sigma_{\hat{\Theta}_2}^2};$$

hence $\hat{\Theta}_1$ is closer than $\hat{\Theta}_2$ if $\sigma_{\hat{\Theta}_1} < \sigma_{\hat{\Theta}_2}$.

7. Given the trivariate normal probability

$$P = \left(\frac{1}{2\pi}\right)^{3/2} \frac{1}{\sigma_x \sigma_y \sigma_z} \iiint_{U^2 \leq B^2} e^{-(1/2) U^2} \, dx \, dy \, dz,$$

where

$$U^2 = \left[\frac{(x - \mu_x)}{\sigma_x}, \frac{(y - \mu_y)}{\sigma_y}, \frac{(z - \mu_z)}{\sigma_z} \right] \begin{bmatrix} 1 & 0 & 0 \\ 0 & 1 & 0 \\ 0 & 0 & 1 \end{bmatrix} \begin{bmatrix} \frac{(x - \mu_x)}{\sigma x} \\ \frac{(y - \mu_y)}{\sigma y} \\ \frac{(z - \mu_z)}{\sigma z} \end{bmatrix},$$

Show that P reduces to

$$P = \frac{1}{\sqrt{2\pi}} \int_\infty^B e^{-(1/2)t^2}\, dt - \frac{2B}{\sqrt{2\pi}} e^{-(1/2)B^2} - 1.$$

8. Given $S = \{1, 2, 3, 4\}$ and the sigma field $R = \{\{1\}, \{2\}, \{1, 2\}, \{3, 4\}, \{1, 3, 4\}, \{2, 3, 4\}, S, 0\}$. Define a function X on S that is a random variable with the elements of R as events. Define real-valued function Y on S that is not a random variable whenever the elements of R are events.

CHAPTER 3

Best Linear Estimation

3-1 Introductory Remarks

In this chapter we formulate a theory to estimate a vector X using information obtained by observing a value of a random vector Y. We restrict ourselves for the most part to linear estimators for X, that is, those estimators of the form

$$\hat{X} = BY,$$

where B is a matrix selected from a class of matrices to ensure that $\hat{X}$ is best in a specified sense.

In applying the theory developed here, one must first determine the stochastic nature of X. That is, is X a random vector or nonrandom vector? If X is random, are its moments known or unknown? Is there a known relation relating X and Y, say,

$$Y = HX,$$

where H is a known mapping matrix? Does one know the density function of X or the joint density function of X and Y?

Basic Principle in Estimation. An estimator may not be optimal with respect to a specified criterion if "all the information" concerning the parameter being estimated is not "used" in modeling the relationship between the observable random vector Y and the parameter X and in formulating the estimate.

The basic Theorem 3-1.1 can be proved easily by using the variation notation introduced in Chapter 1.

THEOREM 3-1.1. *Let* X *and* Y *be real random vectors such that*

$$E(X) = 0 \quad E(XX^T) = V_{XX} \quad E(XY^T) = V_{XY}$$

$$E(Y) = 0 \quad E(YY^T) = V_{YY} \quad \quad\quad\;\; = V_{YX}^T.$$

Then a linear estimator of X *that minimizes the matrix-valued squared error loss function*

$$Q = E[(\hat{X} - X)(\hat{X} - X)^T] \tag{3-1}$$

is

$$\hat{X} = V_{XY} V_{YY}^{-1} Y. \tag{3-2}$$

We say that $\hat{X}$ *is the best linear estimator for* X.

Proof: Since $\hat{X}$ is restricted to be a linear estimator for X, then $\hat{X}$ can be written as

$$\hat{X} = BY, \tag{3-3}$$

and our problem is to select the matrix B so that Q is minimized. On substituting (3-3) into (3-1),

$$Q(B) = E[(BY - X)(BY - X)^T]$$

$$= E[BYY^TB^T - BYX^T - XY^TB^T + XX^T]$$

$$= BV_{YY}B^T - BV_{YX} - V_{XY}B^T + V_{XX}.$$

Then

$$Q(B + \delta B) - Q(B) = \delta B[V_{YY}B^T - V_{YX}]$$

$$+ [BV_{YY} - V_{XY}]\delta B^T + \delta B V_{YY} \delta B^T,$$

which implies

$$\delta Q(B) = \delta B[V_{YY}B^T - V_{YX}] + [BV_{YY} - V_{XY}]\delta B^T.$$

A necessary condition for $\hat{X}$ to be optimal is

$$BV_{YY} - V_{XY} = 0$$

or

$$B = V_{XY}V_{YY}^{-1}. \tag{3-4}$$

This is a sufficient condition for $\hat{X}$ to be optimal, for suppose

$$\hat{X}_1 = B_1 Y$$

is the estimator that minimizes Q. Then $Q(V_{XY}V_{YY}^{-1}) - Q(B_1)$ must be positive semidefinite. But there exists ΔB such that $B_1 = V_{XY}V_{YY}^{-1} + \Delta B$, which implies

$$Q(V_{XY}V_{YY}^{-1}) - Q(B_1) = -\Delta B V_{YY} \Delta B^T.$$

Hence $Q(V_{XY}V_{YY}^{-1}) - Q(B)_1$ is not positive semidefinite. This implies

$$\hat{X} = V_{XY}V_{YY}^{-1}Y$$

is the estimator of X that minimizes Q.

COROLLARY 3-1.1. *If* X *and* Y *are random vectors such that*

$$EX = \mu_X, \quad E[(X - \mu_X)(X - \mu_Y)^T] = V_{XX},$$

$$EY = \mu_Y, \quad E[(Y - \mu_Y)(Y - \mu_Y)^T] = V_{YY},$$

and

$$E[(X - \mu_X)(Y - \mu_Y)] = V_{XY},$$

then

$$\hat{X} = \mu_X + V_{XY}V_{YY}^{-1}(Y - \mu_Y). \tag{3-5}$$

Proof: Let $X_1 = X - \mu_X$ and $Y_1 = Y - \mu_Y$; then $E(X_1) = 0$, $E(Y_1) = 0$, $E(X_1 X_1^T) = V_{XX}$, $E(Y_1 Y_1^T) = V_{YY}$, and $E(X_1 Y_1^T) = V_{XY}$. Then $\hat{X}_1$ follows directly from Theorem 3-1.1. Finally, we note that $\hat{X}_1 = \hat{X} - \mu_X$, since Q is invariant under a linear translation.

COROLLARY 3-1.2. *If* X *and* Y *are random vectors with the moments defined in Corollary 3-1.1 and distributed jointly normal, then* $\hat{X}$ *as defined by* (3-5) *is* $E(X|Y)$, *the Bayesian estimator for* X *with respect to the squared error loss function.*

Proof: See Chapter 2, Sections 2-2 and 2-4.

It is important to note that the above estimators require that the moments of the random vectors X and Y be known. In most applications, this is indeed optimistic, and if one must estimate these moments, the resulting estimators are no longer optimal.

3-2 The Linear Model

It is convenient at times to model the relationship of Y with x by a so-called linear model

$$Y = Hx + U, \qquad (3\text{-}6)$$

where Y is $Np \times 1$ observable random vector, H is $Np \times n$ known mapping matrix, x is an $n \times 1$ state or parameter vector (nonrandom) to be estimated, and U is an *unobservable* $Np \times 1$ vector of random variables.

Example 3-2.1. (Simple Linear Regression)
The equation of a straight line in (t, y) coordinate system is

$$y = x_1 + x_2 t$$

or

$$y = (1, t)\begin{pmatrix} x_1 \\ x_2 \end{pmatrix}. \qquad (3\text{-}7)$$

Suppose we wish to estimate x_1 and x_2, that is, the vector $x = (x_1, x_2)^T$ by by observing N values of Y, say, $y_1, y_2, \ldots, y_N$ at N distinct values of t, ($p = 1$). Then the system of equations

$$y_i = (1, t_i)\begin{pmatrix} x_1 \\ x_2 \end{pmatrix} + u_i, \qquad i = 1, 2, \ldots, N,$$

where u_i is an unknown value of the ith component of the random vector U, can be written in matrix form as in (3-6), where

$$y = (y_1, y_2, \ldots, y_N)^T$$

$$x = (x_1, x_2)^T, \qquad (n = 2)$$

$$H = \begin{bmatrix} 1 & t_1 \\ 1 & t_2 \\ \cdot & \cdot \\ \cdot & \cdot \\ \cdot & \cdot \\ 1 & t_N \end{bmatrix} \quad \text{and} \quad U = \begin{bmatrix} u_1 \\ u_2 \\ \cdot \\ \cdot \\ \cdot \\ u_N \end{bmatrix} = \begin{bmatrix} y_1 - (1, t_1)\begin{pmatrix} x_1 \\ x_2 \end{pmatrix} \\ y_2 - (1, t_2)\begin{pmatrix} x_1 \\ x_2 \end{pmatrix} \\ \cdot \\ \cdot \\ \cdot \\ y_N - (1, t_N)\begin{pmatrix} x_1 \\ x_2 \end{pmatrix} \end{bmatrix}$$

Example 3-2.2. (Linearization of a Nonlinear Model)
Let the $p \times 1$ vector

$$z = f(u, t) = \begin{bmatrix} f_1(u_1, \ldots, u_N, t) \\ f_2(u_1, \ldots, u_N, t) \\ \vdots \\ f_p(u_1, \ldots, u_N, t) \end{bmatrix}$$

be a vector-valued function of the unknown parameter $u = (u_1, \ldots, n_N)^T$ and known variable t. Suppose further that one has an approximation for u, say, u_0, and a Taylor expansion of z about u_0 is well defined. That is, we approximate z by

$$z = f(u^0, t) + \left\{\frac{\partial f_1}{\partial u_j}\right\}_{u_0} (u - u_0) + v,$$

where the $p \times p$ matrix

$$\left\{\frac{\partial f_i}{\partial u_1}\right\}_{u_0} = \begin{bmatrix} \frac{\partial f_1}{\partial u_1}, \ldots, \frac{\partial f_1}{\partial u_N} \\ \vdots \\ \frac{\partial f_p}{\partial u_1}, \ldots, \frac{\partial f_p}{\partial u_N} \end{bmatrix}$$

is evaluated at $u = u_0$, and v is the $p \times 1$ truncation error vector. To estimate the vector u, one obtains N observations of the variable z, at N values of t; then

$$z_i - f(u_0, t_i) + \left\{\frac{\partial f_i(u, t_i)}{\partial u_j}\right\}_{u_0} (u - u_0) + v_i,$$

where the $p \times 1$ vector v_i includes the truncation error v and a random error due to measuring z at t_i. Let

$$Y = \begin{bmatrix} z_1 - f(u_0, t_1) \\ \vdots \\ z_n - f(u_0, t_N) \end{bmatrix}$$

$$x = (u - u_0)$$

$$H = \begin{bmatrix} \dfrac{\partial f_i(u, t_1)}{\partial u_j} \\ \cdot \\ \cdot \\ \cdot \\ \dfrac{\partial f_i(u, t_N)}{\partial u_j} \end{bmatrix}$$

$$v = \begin{bmatrix} v_1 \\ \cdot \\ \cdot \\ \cdot \\ v_n \end{bmatrix}$$

Obviously, the validity of this model will be particularly sensitive to the nonlinearity of $f(u, t)$ and the degree of accuracy to which u_0 approximates u. When an estimate for x is obtained, the estimate for u follows from

$$\hat{u} = \hat{x} + u_0,$$

where $\hat{x}$ is the estimate for x.

3-3 The Classical Form of the Gauss–Markov Theorem

The proof presented here for the Gauss–Markov theorem is the one usually found in most statistics texts.

THEOREM 3-3.1. (*Gauss–Markov Theorem*). Let

$$Y = Hx + U \tag{3-8}$$

be a linear model, where Y *is an* $Np \times 1$ *observable random vector,* H *is an* $Np \times n$ *known mapping matrix of rank* $n \leq Np$, x *is an* $n \times 1$ *unknown nonrandom parameter (state) vector, and* U *is an* $Np \times 1$ *vector whose elements are random such that*

$$E(U) = 0,$$

and

$$E(UU^T) = V,$$

a known positive definite covariance matrix. Then the minimum variance linear unbiased estimator of x, *denoted by* $\hat{X}$, *is given by*

Sec. 3-3 The Classical Form of the Gauss–Markov Theorem

$$\hat{X} = (H^TV^{-1}H)^{-1}H^TV^{-1}Y \tag{3-9}$$

whose variance $V_{\hat{x}}$ is given by

$$V_{\hat{x}} = (H^TV^{-1}H)^{-1}. \tag{3-10}$$

Proof: Let X^* be any linear unbiased estimator that is minimal with respect to the matrix covariance denoted by V_{X^*}.

$$V_{X^*} = E[(X^* - x)(X^* - x)^T],$$

where

$$X^* = BY \quad \text{(linearity property)}.$$

By the unbiased requirement,

$$EX^* = E[B(Hx + U)] = BHx = x$$

imposes the constraint

$$(BH - I)x = 0$$

for every x, or

$$BH = I. \tag{3-11}$$

Let $B = (H^TV^{-1}H)^{-1}H^TV^{-1} + C$, where C is simply a matrix difference. Thus it follows that

$$BH = (H^TV^{-1}H)^{-1}H^TV^{-1}H + CH$$
$$= I + CH,$$

which by (3-11) implies that

$$CH = 0. \tag{3-12}$$

Hence

$$\begin{aligned}
V_{X^*} &= E[(BY - x)(BY - x)^T] \\
&= E[(H^TV^{-1}H)^{-1}H^TV^{-1}Y + CY - x] \\
&\quad \times [(H^TV^{-1}H)^{-1}H^TV^{-1}Y + CY - x]^T \\
&= E[(\hat{X} - x + CY)(\hat{X} - x + CY)^T] \\
&= E[(\hat{X} - x)(\hat{X} - x)^T] - E[CY(\hat{X} - x)^T] \\
&\quad - E[(\hat{X} - x)Y^TC] + E[CYY^TC].
\end{aligned} \tag{3-13}$$

We can write the center terms in (3-13) as

$$E[CY(\hat{X} - x)^T] = E\{(CY)[(H^TV^{-1}H)^{-1}H^TV^{-1}Y - x]^T\}.$$

By substituting (3-8) one obtains

$$E[CY(\hat{X} - x)^T] = E\{C(Hx + U)[(H^TV^{-1}H)^{-1}H^TV^{-1}(Hx + U) - x]^T\}$$

$$= E\{CU[x + (H^TV^{-1}H)^{-1}H^TV^{-1}U - x]^T\}$$

$$= [CE(UU^T)V^{-1}H(H^TV^{-1}H)^{-1}]$$

$$= [CVV^{-1}H(H^TV^{-1}H)^{-1}]$$

$$= [CH(H^TV^{-1}H)^{-1}]$$

$$= 0,$$

since $CH = 0$ by (3-12).

It should be noted that if $\hat{X}$ is minimal in the sense of minimum mean-square error, then

$$E[(\hat{X} - x)Y^T] = 0.$$

This is related to an *orthogonality principle* well known in engineering sciences. This principle will be discussed later in Section 3-8.

It follows then that

$$V_{X^*} = V_{\hat{x}} + E\{C[YY^T]C^T\}$$

$$= V_{\hat{x}} + CE[YY^T]C^T$$

or

$$V_{X^*} - V_{\hat{x}} = CE[YY^T]C^T,$$

a positive definite or semidefinite matrix. Since $E[YY^T]$ is positive definite, it follows that in order for $V_{X^*} = V_{\hat{x}}$, then

$$CE[YY^T]C^T = 0,$$

which can happen if and only if

$$C = 0.$$

This in turn implies that $X^* = \hat{X}$ if and only if $\hat{X}$ is given by (3-9). The covari-

Sec. 3-3 The Classical Form of the Gauss–Markov Theorem

ance of $\hat{X}$ follows easily from Definition 2-3.3 and (3-8).

$$V_{\hat{x}} = E[(\hat{X} - x)(\hat{X} - x)^T]$$

$$= E[(H^TV^{-1}H)^{-1}H^TV^{-1}UU^TV^{-1}H(H^TV^{-1}H)^{-1}]$$

$$= (H^TV^{-1}H)^{-1}H^TV^{-1}VV^{-1}H(H^TV^{-1}H)$$

$$= (H^TV^{-1}H)^{-1}.$$

Theorem 3-3.1 is proved.

COROLLARY 3-3.1. *Let* $x' = p^Tx$, *where* p *is a* n $\times$ 1 *vector of scalars; then the best linear unbiased estimator (BLUE) of* p^Tx *is simply* $p^T\hat{X}$, *where* $\hat{X}$ *is given by* (3-9).

COROLLARY 3-3.2. *Let* $x' = $ Px, *where* P *is* k $\times$ n *and rank of* P *is* k; *then the best linear unbiased estimator of* Px *is* P$\hat{X}$.

COROLLARY 3-3.3. *Suppose that the vector* U *is distributed according a multivariate normal density function with mean zero and covariance matrix* V; *then* $\hat{X}$ *is distributed multivariate normal with mean* x *and covariance* $(H^TV^{-1}H)^{-1}$.

Proof: This fact follows directly from Theorem 2-3.1.

COROLLARY 3-3.4. *If* $V = \sigma^2 I$, *then*

$$\hat{X} = (H^TH)^{-1}H^TY \tag{3-14}$$

is not only BLUE but is also the estimator that minimizes the sum of squared deviations, the so-called least square estimate, (p = 1).

Proof: Consider the deviation

$$U_i = Y_i - h_i x, \quad i = 1, 2, \ldots, N,$$

where U_i is the deviation and h_i is the ith row of the matrix H. Then the sum of the squared deviations

$$\sum_{i=1}^{N} U_i^2$$

can be written

$$U^T U = (Y - Hx)^T(Y - Hx).$$

A necessary condition for $U^T U$ to be minimum is

$$\frac{d(U^T U)}{dx} = 0,$$

or

$$\frac{d(Y^T Y - 2x^T H^T Y + x^T H^T H x)}{dx} = 0,$$

which implies

$$-2H^T Y + 2H^T H x = 0;$$

that is,

$$H^T H x = H^T Y,$$

the so-called *normal equations*. We note that the rank of H is n; hence rank $H^T H$ of an $n \times n$ matrix is n. This in turn implies that $(H^T H)^{-1}$ exists, and the least square estimator

$$X_{LS} = (H^T H)^{-1} H^T Y$$

is indeed the same as (3-14).

It is important to note that even if σ^2 is unknown, (3-14) yields the BLUE; however, V must be known if (3-9) is to be used. A very complicated problem exists if V is unknown and known not to be of the form $\sigma^2 I$.

By using the properties of the pseudoinverse of a matrix, an extension of the Gauss–Markov theorem is possible [119].

THEOREM 3-3.2.* *Let the linear model be defined as*

$$Y = Hx + U; \quad EU = 0; \quad EUU^T = V,$$

where Y *is a* p × 1 *vector, x is an* n × 1, H *is a* p × n *matrix, and* U *is a* p × 1 *random vector. Suppose that rank of the matrix* H *is* r, *then the linear estimator that minimizes* $E[(\hat{X} - x)(\hat{X} - x)]^T$ *and* $E(\hat{X}) = x$ *if x is in the range space of* H^T, *is given by*

$$\hat{X} = M^+ H^T V^{-1} Y$$

$$M = H^T V^{-1} H$$

---

* This theorem is included for completeness mathematically and lends little to the discussion; it is suggested that the reader read the proof after he reads the remainder of the chapter.

Sec. 3-3 The Classical Form of the Gauss–Markov Theorem

and

$$V_{\hat{X}} = M^+.$$

Proof: We require that $\hat{X} = BY$ and $E(\hat{X}) = x$ whenever $x \in R(H^T)$. These requirements imply that $E(\hat{X}) = BHx$ and the property $R(H^+) = R(H^T)$ implies that for x in $R(H^T)$,

$$H^+Hx = BHx = x.$$

Let $x = x_1 + x_2$, where $x_1 \in R(H^T)$, and $x_2 \in N(H)$. Then

$$\|E(\hat{X}) - x\| = \|BHx_1 + BHx_2 - x\| = \|BHx_1 - x\| = \|x_2\|.$$

Thus it follows that $\|E(\hat{X}) - x\|$ is minimum with respect to $R(H^T)$ for $x \notin R(H^T)$. The covariance matrix $V_{\hat{X}}$ of the estimator $\hat{X}$ is given by $V_{\hat{X}} = BVB^T$ and must be minimized subject to the constraint $BH = H^+H$. To do this we adjoin the constraint $BH = H^+H$ to BVB^T using a matrix Lagrangian multiplier λ and find conditions necessary to minimize

$$Q = BVB^T + \lambda^T[H^+H - H^TB^T] + [H^+H - BH]\lambda.$$

Employing the variational technique we obtain the first variation δQ,

$$\delta Q = \delta B[VB^T - H\lambda] + [BV - \lambda^TH^T]\delta B^T.$$

Since δB is arbitrary, we find that setting $\delta Q = \phi$ implies

$$BV - \lambda^TH^T = \phi$$

or

$$B = \lambda^TH^TV^{-1}.$$

Multiplying the latter by H we obtain

$$H^+H = \lambda^TH^TV^{-1}H.$$

Therefore, if we set $H^TV^{-1}H = M$, we have

$$\lambda^T = M^+ + z(I - MM^+),$$

where z is arbitrary to within having the dimension of λ^T. To see this we need to show that

$$H^+HM^+M = H^+H.$$

$$H^+H(H^TV^{-1}H)^+(H^TV^{-1}H) = H^+H.$$

It follows that $R(M^+) = R(M^T) = R(M)$. Hence we must show that $R(M) = R(H^T)$. We observe that $R(H^T) \supset R(M)$ and $N(H) \subset N(M)$. Suppose there exists $x \in N(M)$ such that $x \notin N(H)$. Then $Hx \neq 0$ and $Mx = 0$. But since V^{-1} is positive definite, $x^T M x \neq 0$, which implies $Mx \neq 0$. This is a contradiction. Hence $N(H) = N(M)$. Since $R(M^T)$ and $N(M)$ are orthogonal spaces and their sum is the n-dimensional vector space X_n, it follows that $R(M) = R(H^T)$. We now observe that the columns of H^+H are in $R(M^+)$; thus $M^+MH^+H = H^+H$. Taking the transpose of both sides gives

$$H^+HM^+M = H^+H.$$

Assume that the rank of H is $q \leq \min(n, p)$. Then

$$B = \lambda^T H^T V^{-1}$$

$$= (M^+ + z[I - MM^+])H^T V^{-1}$$

$$= M^+ H^T V^{-1} + z[I - MM^+]H^T V^{-1}.$$

To establish that the second term is ϕ, that is, $(I - MM^+)H^T V^{-1} = \phi$, we observe that $(I - MM^+)$ is an orthogonal projection on the null space of M^+. We need to show that $N(M^+) = N(H)$. Since $M = H^T V^{-1} H$, then it certainly follows that $N(M) = N(M^T)$. Also note that $N(M) \supseteq N(H)$. Thus, suppose there exists an $x \in N(M)$ such that $x \notin N(H)$. Hence it follows that $Hx \in R(H)$. Since V^{-1} is positive definite, V^{-1} does not rotate Hx into the null space of H^T. Hence $H^T V^{-1} Hx \neq 0$, which implies $x \notin N(M)$. This is a contradiction. Thus $N(M) = N(M^T)$. Now $N(M) = N(M^T) = N(M^+)$, which implies $N(M^+) = N(H)$, and consequently, $(I - MM^+)H^T V^{-1} = \phi$, since $R(H^T) = N(H)^\perp$, where $(\cdot)^\perp$ denotes the orthogonal complement of $(\cdot)$. Therefore,

$$\hat{X} = BY = M^+ H^T V^{-1} Y,$$

and the covariance matrix

$$V_{\hat{X}} = BVB^T = M^+ M M^{+T} = M^+,$$

which are the desired results.

It should be noted that if the rank of H is equal to $n \leq p$, then $H^+H = I$ and $\hat{X}$ is $(H^T V^{-1} H)^{-1} H^T V^{-1} Y$ and its covariance matrix is given by $(H^T V^{-1} H)^{-1}$. If the rank of H is $p < n$, then $HH^+ = I$, $(H^T V^{-1})^+ = V H^{T+}$, $\hat{X} = H^+ Y$, and $V_{\hat{X}} = H^+ V H^{+T}$.

3-4 Comparison of Least Squares and Minimum Variance Estimators

It is of interest to compare the least squares estimator of the state vector to that of the minimum variance estimator of the state vector. Magness and McGuire [125] have been able to give an extensive analysis in comparing these two estimates whenever the mapping matrix H of the linear model is of full rank (columns linearly independent). They were able to establish the inequality

$$V_{LS} \leq \frac{1}{4}(\lambda_{max} + \lambda_{min})\left(\frac{1}{\lambda_{max}} + \frac{1}{\lambda_{min}}\right)V_{MV},$$

where V_{LS} and $V_{MV} = (\sigma_{ij})$ are the covariance matrices of the least squares estimator and minimum variance estimator, respectively. λ_{max} and λ_{min} are the maximum and minimum eigenvalues of the correlation matrix $\rho = \{\sigma_{ij}/(\sigma_{ii}\sigma_{jj})^{1/2}\}$ of the error vector. The above inequality places an upper bound on how much is lost by use of the least squares estimator of the state vector to that of the minimum variance estimator of the state vector.

In Theorem 3-4.1 it will be shown that the least squares estimator of the state vector will have the same covariance matrix as that of the best linear estimator of the state vector, whenever the mapping matrix of the linear model has all of its rows linearly independent.

THEOREM 3-4.1. Consider the linear model described by the vector equation in Theorem 3-3.2, where the rank of H is p. Then the covariance matrix of the least squares estimator of the state vectors equals the covariance of the best linear estimator of the state vector.

Proof: The least squares estimator of the state vector is

$$\hat{X}_{LS} = (H^T H)^+ H^T Y$$

$$= H^+ Y.$$

The corresponding covariance matrix is

$$V_{LS} = H^+ V H^{T+}.$$

The best linear estimator of the state vector is by Theorem 3-3.2:

$$\hat{X} = H^+ Y.$$

The corresponding covariance matrix is

$$V_{\hat{x}} = H^+ V H^{T+}.$$

Thus it can be seen that there is no loss in using the least squares estimator whenever the rows of the mapping matrix are linearly independent.

3-5 The Recursive Form of the Estimator

In real-time estimation problems (filtering) it is necessary that the estimator $\hat{X}$ be written in a recursive form. This can be done easily by the following formulation:

$$\hat{X}_N = \hat{X}_{N-1} - \hat{X}_{N-1} + [H_N V_N^{-1} H_N] H_N V_N^{-1} Z_N,$$

where $\hat{X}_N$ is the best estimator given $Np \times 1$ data vectors Y_i, where

$$Z_N = \begin{bmatrix} Y_1 \\ Y_2 \\ \cdot \\ \cdot \\ \cdot \\ Y_N \end{bmatrix}$$

is an $Np \times 1$ vector of observable random variables, where

$$U_N = \begin{bmatrix} D_1 \\ D_2 \\ \cdot \\ \cdot \\ \cdot \\ D_N \end{bmatrix}$$

is an $Np \times 1$ error vector, and where

$$H_N = \begin{bmatrix} h_1 \\ h_2 \\ \cdot \\ \cdot \\ \cdot \\ h_N \end{bmatrix}$$

is an $Np \times n$ mapping matrix relating Z_N; x, the parameter vector; and

Sec. 3-5 The Recursive Form of the Estimator

U_N in the linear model

$$Z_N = H_N x + U_N.$$

Assuming that the covariance matrix of U_N is block diagonal, that is,

$$V_N = \text{Cov}(U_N) = \begin{bmatrix} V_{11} & \phi & \cdots & \phi \\ \phi & V_{22} & \cdots & \phi \\ \cdot & \cdot & \cdot & \cdot \\ \cdot & \cdot & \cdot & \cdot \\ \cdot & \cdot & \cdot & \cdot \\ \phi & \phi & \cdots & V_{NN} \end{bmatrix},$$

then

$$\hat{X}_N = \hat{X}_{N-1} - \hat{X}_{N-1} + [H_{N-1}^T V_{N-1}^{-1} H_{N-1} + h_N^T V_{NN}^{-1} h_N]^{-1}$$
$$\times [H_{N-1}^T V_{N-1}^{-1} Z_{N-1} + h_N^T V_{NN}^{-1} Y_N]$$

or

$$\hat{X}_N = \hat{X}_{N-1} + [H_{N-1}^T V_{N-1}^{-1} H_{N-1} + h_N^T V_{NN} h_N]^{-1} [H_{N-1}^T V_{N-1}^{-1} Z_{N-1}$$
$$+ h_N^T V_{NN}^{-1} Y_N - H_{N-1}^T V_{N-1}^{-1} H_{N-1} \hat{X}_{N-1} - h_N^T V_{NN}^{-1} h_N \hat{X}_{N-1}],$$

which reduces to

$$\hat{X}_N = \hat{X}_{N-1} + [H_{N-1}^T V_{N-1}^{-1} H_{N-1} + h_N^T V_{NN}^{-1} h_N]^{-1} h_N^T V_{NN}^{-1} [Y_N - h_N \hat{X}_{N-1}],$$

since

$$\hat{X}_{N-1} = [H_{N-1}^T V_{N-1}^{-1} H_{N-1}]^{-1} H_{N-1}^T V_{N-1}^{-1} Z_{N-1}.$$

A recursive form for the Cov $\hat{X}_N$ can be obtained using the "inside-out" rule for inverting the sum of a positive semidefinite matrix and a positive definite matrix (see Theorem 1-4.9). That is,

$$\text{Cov } \hat{X}_N = [H_N^T V_N^{-1} H_N]^{-1}$$
$$= [H_{N-1}^T V_{N-1}^{-1} H_{N-1} + h_N^T V_{NN}^{-1} h_N]^{-1}$$
$$= (H_{N-1}^T V_{N-1}^{-1} H_{N-1})^{-1} - (H_{N-1}^T V_{N-1}^{-1} H_{N-1})^{-1} h_N^T [V_{NN}$$
$$+ h_N (H_{N-1}^T V_{N-1}^{-1} H_{N-1})^{-1} h_N^T]^{-1} h_N (H_{N-1}^T V_{N-1}^{-1} H_{N-1})^{-1}$$
$$= \text{Cov}(\hat{X}_{N-1}) - \text{Cov}(\hat{X}_{N-1}) h_N^T [V_{NN}$$
$$+ h_N \text{Cov}(\hat{X}_{N-1}) h_N^T]^{-1} h_N \text{Cov}(\hat{X}_{N-1}),$$

which gives a recursive way for computing Cov $(\hat{X}_N)$ as a function of the Cov $(\hat{X}_{N-1})$.

This technique allows one to invert large covariance matrices by simply inverting one of a smaller dimension.

3-6 The Gauss–Markov Theorem When the Parameter Vector is Random

In Section 3-3 the parameter vector x was assumed to be a constant vector. There are cases when it is natural to assume that the vector X in the linear model

$$Y = HX + U \tag{3-15}$$

is random with the following statistical properties:

$$E(X) = \mu_X \tag{3-16}$$

$$E[(X - \mu_X)(X - \mu_X)^T] = V_{XX}. \tag{3-17}$$

The vector U is a $p \times 1$ vector of random elements such that

$$E(U) = \phi \tag{3-18}$$

$$E(UU^T) = V \tag{3-19}$$

$$E(UX^T) = \phi. \tag{3-20}$$

From (3-16) through (3-20) it follows easily that

$$\begin{aligned}
E(Y) &= H\mu_X \\
V_{YY} &= E[(Y - H\mu_X)(Y - H\mu_X)^T] \\
&= E[(HX + U - H\mu_X)(HX + U - H\mu_X)^T] \\
&= E[H(X - \mu_X)(X - \mu_X)^T H^T + U(X - \mu_X)^T H^T \\
&\quad + H(X - \mu_X)U^T + UU^T] \\
V_{YY} &= HV_{XX}H^T + V.
\end{aligned} \tag{3-21}$$

Also, $V_{XY} = E[(X - \mu_X)(Y - \mu_Y)^T] = E[(X - \mu_X)(X - \mu_X)^T H^T + (X - \mu_X)U^T]$
$= V_{XX}H^T$.

Sec. 3-6   Random Parameter Vector

We shall consider the class of estimators defined by the formula

$$\hat{X} = a + AY,$$

where a is a vector of real numbers, and A is a matrix defined on the real numbers selected so that

$$E[\hat{X} - X] = \phi \qquad (3\text{-}22)$$

and

$$Q = E[(\hat{X} - X)(\hat{X} - X)^T]$$

is minimized in the usual sense.

Consider the constraint (3-22):

$$\phi \equiv E(\hat{X} - X) = E[a + AY - X]$$

or

$$\phi = a + AH\mu_X - \mu_X.$$

We first consider the case when μ_X is known.

THEOREM 3-6.1. *Let* $a + AY$ *be a linear estimator of* X *in the linear model* (3-15). *Then the optimum values of* a *and* A *for which*

$$E(a + AY - X)(a + AY - X)^T$$

is a minimum are

$$a^* = \mu_X - A^*H\mu_X \qquad (3\text{-}23)$$

$$A^* = V_{XX}H^T(HV_{XX}H^T + V)^{-1}. \qquad (3\text{-}24)$$

The variance of the estimator is

$$V_{XX}H^T[H^TV_{XX}H + V]^{-1}HV_{XX}.$$

Proof: Let $\hat{X} = a + AY$ be an estimator for X. Then consider

$$Q = E[(\hat{X} - X)(\hat{X} - X)^T]$$
$$= E[(a + AY - X)(a + AY - X)^T]$$
$$= aa^T + a\mu_Y^T A^T - a\mu_X^T + A\mu_Y a^T + A[V_{YY} + \mu_Y \mu_Y^T]A^T$$
$$\quad - A[V_{YX} + \mu_Y \mu_X^T] - \mu_X a^T - [V_{XY} + \mu_X \mu_Y^T]A^T + [V_{XX} + \mu_X \mu_X^T].$$

A necessary condition for Q to be minimal is for the first variations in Q with respect to a and the first variation in Q with respect to A to be simultaneously the null vector and the null matrix, respectively. Let $\delta_a Q$ and $\delta_A Q$ denote these variations. Hence

$$\delta_a Q = \delta a [a^T + \mu_Y^T A^T - \mu_X^T] + [a + A\mu_Y - \mu_X]\delta a^T$$

$$\delta_A Q = \delta A[\mu_Y a^T + (V_{YY} + \mu_Y \mu_Y^T)A^T - (V_{YX} + \mu_Y \mu_X^T)]$$
$$+ [a\mu_Y^T + A(V_{YY} + \mu_Y \mu_Y^T) - (V_{XY} + \mu_X \mu_Y^T)]\delta A^T.$$

The constraints $\delta_a Q \equiv \phi$ and $\delta_A Q \equiv \phi$ for all δa and δA, respectively, imply that

$$a^* + A^*\mu_Y - \mu_X \equiv \phi$$

and $a^*\mu_Y^T + A^*(V_{YY} + \mu_Y \mu_Y^T) - (V_{XY} + \mu_X \mu_X^T) = \phi$. The first condition implies $a^* = \mu_X - A^*\mu_Y$. Since $\mu_Y = h\mu_X$, it follows that

$$a^* = \mu_X - A^*h\mu_X.$$

The second condition implies that

$$A^* = V_{XY}V_{YY}^{-1} = V_{XX}h^T[hV_{XX}h^T + V]^{-1},$$

the desired results. We note that

$$\hat{X} = \mu_X + V_{XY}V_{YY}^{-1}(Y - h\mu_X). \tag{3-25}$$

Note that this result is that in Corollary 3-1.1 with $EY = h\mu_X$.

The covariance matrix, $V_{\hat{X}}$, for $\hat{X}$ follows easily from the definition

$$V_{\hat{X}} = V_{XY}V_{YY}^{-1}V_{YY}V_{YY}^{-1}V_{YX}$$
$$= V_{XY}V_{YY}^{-1}V_{YX}$$
$$= V_{XX}h^T[h^T V_{XX}^{-1}h + V]^{-1}hV_{XX}.$$

The proof is complete.

It is clear from (3-23) or (3-25) that if μ_X is not known, then the estimator (3-25) is not computable. We now consider the case where μ_X is not known.

THEOREM 3-6.2. *Suppose that* a *and* μ_X *are independent mathematically and* μ_X *is unknown. Then the optimal values of* a *and* A *for which*

1. $E[a + AY - X] = \phi$ *for all values of* μ_X

2. $E[(a + AY - X)(a + AY - X)^T]$ *is a minimum are*

$$a^* = \phi$$

$$A = (H^T V^{-1} H)^{-1} H^T V^{-1}.$$

The covariance matrix of the estimator is $(H^T V^{-1} H)^{-1}$.

Proof: The first condition implies that

$$a + AH\mu_x - \mu_x = \phi$$

for all μ_x. This in turn implies that

$$a + (AH - I)\mu_x \equiv \phi$$

for all μ_x. Since a and μ_x are mathematically independent, then

$$a = \phi$$

$$AH - I = \phi,$$

or we select A so that

$$AH = I.$$

Let

$$Q = E[(AY - X)(AY - X)^T + \lambda^T[I - H^T A^T] + [I - AH]\lambda, \quad (3\text{-}26)$$

where λ is a matrix Lagrangian multiplier. Substituting $AH = I$ and $Y = HX + U$ into (3-26) we get

$$Q = AVA^T + \lambda^T[I - H^T A^T] + [I - AH]\lambda.$$

Equating the first variation of Q with respect to A to zero, one obtains

$$AV - \lambda^T H^T = \phi$$

$$AV = \lambda^T H^T \quad (3\text{-}27)$$

$$A = \lambda^T H^T V^{-1}.$$

Multiplying both sides of (3-27) by H on the right:

$$I = \lambda^T H^T V^{-1} H.$$

From (3-27), it follows that

$$A = (H^T V^{-1} H)^{-1} H^T V^{-1}.$$

3-7 On Estimating a Subvector of X

The minimum variance linear unbiased estimator for nonrandom x in the linear model

$$Y = Hx + v$$

is

$$\hat{X} = (H^T V^{-1} H)^{-1} H^T V^{-1} Y. \tag{3-28}$$

Let x be partitioned such that

$$x = \begin{bmatrix} x^{(1)} \\ x^{(2)} \end{bmatrix},$$

where $x^{(1)}$ is a $q \times 1$ vector and $x^{(2)}$ is an $n - q \times 1$ vector. Suppose we wish to estimate $x^{(1)}$ by a minimum variance estimator and yet not estimate the total vector x by using (3-28).

THEOREM 3-7.1. The minimum variance linear unbiased estimator for $x^{(1)}$, the first q elements of x, is given by the vector

$$\hat{X}^{(1)} = [H_1^T(V^{-1} - V^{-1}H_2\{H_2^T V^{-1} H_2\}^{-1} H_2^T V^{-1})H_1]^{-1} H_1^T(V^{-1}$$

$$- V^{-1}H_2\{H_2^T V^{-1} H_2\}^{-1} H_2^T V^{-1})Y,$$

where H is partitioned such that $H = (H_1 H_2)$ with H_i and $x^{(i)}$ $i = 1, 2$, have dimensions that are compatible for matrix multiplication.

Proof: By partitioning the covariance matrix of $\hat{X}$ we can write the covariance matrix $V_{\hat{X}^{(1)}}$ of $\hat{X}^{(1)}$ as

$$V_{\hat{X}} = [(H_1 H_2)^T V^{-1}(H_1 H_2)]^{-1} = \begin{bmatrix} H_1^T V^{-1} H_1 & H_1^T V^{-1} H_2 \\ H_2^T V^{-1} H_1 & H_2^T V^{-1} H_2 \end{bmatrix}^{-1}. \tag{3-29}$$

Inverting the matrix (3-29), we find that

$$V_{\hat{X}^{(1)}} = [H_1^T V^{-1} H_1 - H_1^T V^{-1} H_2 (H_2^T V^{-1} H_2)^{-1} H_2^T V^{-1} H_1]^{-1}$$

$$= [H_1^T(V^{-1} - V^{-1}H_2\{H_2^T V^{-1} H_2\}^{-1} H_2^T V^{-1})H_1]^{-1}.$$

Following the form of (3-28) we write,

Sec. 3-8   The Orthogonality Principle

$$\hat{X}^{(1)} = [H_1^T(V^{-1} - V^{-1}H_2\{H_2^TV^{-1}H_2\}^{-1}H_2^TV^{-1})H_1]^{-1}$$
$$\times H_1^T(V^{-1} - V^{-1}H_2\{H_2^TV^{-1}H_2\}^{-1}H_2^TV^{-1})Y.$$

On taking the expectation of $\hat{X}^{(1)}$,

$$E\hat{X}^{(1)} = [H_1^T(V^{-1} - V^{-1}H_2\{H_2^TV^{-1}H_2\}^{-1}H_2^TV^{-1}H_1]^{-1}$$
$$\times H_1^T(V^{-1} - V^{-1}H_2\{H_2^TV^{-1}H_2\}^{-1}H_2^TV^{-1})(H_1 H_2)\begin{bmatrix}x^{(1)}\\x^{(2)}\end{bmatrix}$$
$$= x^{(1)}$$

since

$$(V^{-1} - V^{-1}H_2\{H_2^TV^{-1}H_2\}^{-1}H_2^TV^{-1})H_2 x^{(2)} = \phi$$

The fact that $\hat{X}$ is minimum variance implies that indeed $\hat{X}^{(1)}$ is minimum variance.

This partitioning can help in eliminating the necessity of computing unwanted parameters, which may represent sources of systematic errors.

3-8 The Orthogonality Principle

A principle that has proved extremely useful to the engineer can be summarized in Theorem 3-8.1.

THEOREM 3-8.1. Let X and Y *be random variables. Then the matrix* L *in the linear estimate* $\hat{X} = LY$ *of* X *that minimizes the mean squared error*

$$E[(\hat{X} - X)(\hat{X} - X)^T]$$

is such that $\hat{X} - X$ *is orthogonal to* Y; *that is,*

$$E[(\hat{X} - X)Y^T] = 0. \qquad (3\text{-}30)$$

Proof: Let $X^* = L_1 Y$ be the linear estimator of X that minimizes the mean-square error functions; that is, suppose

$$E[(X^* - X)(X^* - X)^T]$$

is a minimum. Now let $\hat{X} = LY$ satisfy (3-30). Define L_2 such that

$$L_1 = L + L_2.$$

Then

$$E[(X^* - X)(X^* - X)^T] = E[(L_1 Y - X)(L_1 Y - X)^T]$$
$$= E[(LY + L_2 Y - X)(LY + L_2 Y - X)^T]$$
$$= E[(LY - X)(LY - X)^T] - EL_2 Y(LY - X)^T$$
$$\quad - E(LY - X)Y^T L_2^T + EL_2 YY^T L_2^T$$
$$= E[(\hat{X} - X)(\hat{X} - X)^T] - L_2 EY(\hat{X} - X)^T$$
$$\quad - E(\hat{X} - X)Y^T L_2^T + L_2 EYY^T L_2^T$$
$$= E[(\hat{X} - X)(\hat{X} - X)^T] + L_2 EYY^T L_2^T.$$

Since X^* is minimal if and only if $L_2 = 0$, equivalently if and only if $X^* = \hat{X}$, thus the theorem is proved.

COROLLARY 3-8.1. *The minimum mean-square error is given by*

$$E[(X - LY)X^T].$$

Proof:

$$E[(LY - X)(LY - X)^T] = E[(LY - X)Y^T]L^T - E[(LY - X)X^T]$$
$$= E[(X - LY)X^T],$$

since $E[(LY - X)Y^T] = \phi$.

EXAMPLE 3-8.1. Let X and Y be random vectors such that

$$E(X) = 0, \quad E(Y) = 0, \quad E(XX^T) = V_{XX},$$
$$E(YY^T) = V_{YY}, \quad \text{and } E(XY^T) = V_{XY}.$$

The formula for $\hat{X}$ follows from (3-30) since

$$\hat{X} = LY$$

implies

$$0 = E[(\hat{X} - X)Y^T] = E[(LY - X)Y^T]$$
$$= LE(YY^T) - E(XY^T)$$
$$= LV_{YY} - V_{XY}.$$

This implies that

$$L = V_{XY}V_{YY}^{-1}$$

and

$$\hat{X} = V_{XY}V_{YY}^{-1} Y,$$

which is identical to (3-2) obtained previously.

What is important to note here is that L need only be a linear operator that commutes with the expectation operator, and for which L_2 exists such that $L_1 = L + L_2$ has meaning. An integral operator as well as a sum has this property.

3-9 On Combining Estimators and Observations

There are cases in which one has available two unbiased estimators, say, $\hat{X}_1$ and $\hat{X}_2$, which estimate the nonrandom parameter x. Usually, one must be content with an estimator of the form

$$\hat{X} = B_1\hat{X}_1 + B_2\hat{X}_2, \qquad (3\text{-}31)$$

where B_1 and B_2 are matrices selected so that

$$E\hat{X} = x \qquad (3\text{-}32)$$

and

$$Q = E[(\hat{X} - x)(\hat{X} - x)^T] \qquad (3\text{-}33)$$

is minimized. Let

$$E[(\hat{X}_1 - x)(\hat{X}_1 - x)^T] = V_{11}$$

$$E[(\hat{X}_2 - x)(\hat{X}_2 - x)^T] = V_{22}$$

and

$$E[(\hat{X}_1 - x)(\hat{X}_2 - x)^T] = V_{12} = V_{21}^T.$$

THEOREM 3-9.1. *The best estimate $\hat{X}$ for x, such that (3-31) and (3-32) are satisfied and (3-33) is minimized, is given by (3-31), when*

$$B_1 = (V_{22} - V_{21})(V_{11} + V_{22} - V_{21} - V_{12})^{-1} \qquad (3\text{-}34)$$

$$B_2 = (V_{11} - V_{12})(V_{11} + V_{22} - V_{21} - V_{12})^{-1} \qquad (3\text{-}35)$$

Proof: Let $\hat{X} = B_1\hat{X}_1 + B_2\hat{X}_2$ be minimal with respect to Q. The condition

$$x = E(\hat{X}) = E(B_1\hat{X}_1 + B_2\hat{X}_2) = B_1 E(\hat{X}_1) + B_2 E(\hat{X}_2) = B_1 x + B_2 x$$
$$= (B_1 + B_2)x$$

for all x, implies

$$B_1 + B_2 = I$$

or

$$B_2 = I - B_1.$$

Then

$$Q = E[\{B_1\hat{X}_1 + (I - B_1)\hat{X}_2 - B_1 x - (I - B_1)x\}$$
$$\times \{B_1\hat{X}_1 + (I - B_1)\hat{X}_2 - B_1 x - (I - B_1)x\}^T]$$
$$= B_1 E(\hat{X}_1 - x)(\hat{X}_1 - x)^T B_1^T + B_1 E(\hat{X}_1 - x)(\hat{X}_2 - x)^T (I - B_1)^T$$
$$+ (I - B_1) E(\hat{X}_2 - x)(\hat{X}_1 - x)^T B_1^T$$
$$+ (I - B_1) E(\hat{X}_2 - x)(\hat{X}_2 - x)^T (I - B_1).$$

On performing the expectation operation,

$$Q = B_1 V_{11} B_1^T + B_1 V_{12}(I - B_1)^T (I - B_1) V_{21} B_1^T$$
$$+ (I - B_1) V_{22}(I - B_1)^T$$
$$= B_1 V_{11} B_1^T + B_1 V_{12} - B_1 V_{12} B_1^T + V_{21} B_1^T - B_1 V_{21} B_1^T$$
$$+ (V_{22} - B_1 V_{22} - V_{22} B_1^T + B_1 V_{22} B_1^T)$$
$$= B_1[V_{11} + V_{22} - V_{12} - V_{21}]B_1^T - B_1^T[V_{22} - V_{12}]$$
$$- [V_{22} - V_{21}]B_1^T + V_{22}.$$

Setting the first variation of Q with respect to B_1 equal to zero, that is,

$$\delta_{B_1} Q = \delta B_1[(V_{11} + V_{22} - V_{12} - V_{21})B_1^T - (V_{22} - V_{12})]$$
$$+ [B_1(V_{11} + V_{22} - V_{12} - V_{21}) - (V_{22} - V_{21})]\delta B_1^T$$
$$= 0$$

Sec. 3-9 On Combining Estimators and Observations

for every δB_1 implies that

$$B_1(V_{11} + V_{22} - V_{12} - V_{21}) - (V_{22} - V_{21}) = 0$$

or B_1 is given by (3-34). The formula (3-35) can be obtained in a similar fashion.

COROLLARY 3-9.1. *Let Y_N be a $p \times 1$ observable vector and let Y_N be related to the $n \times 1$ nonrandom parameter vector x by the linear model*

$$Y_N = H_N x + U_N,$$

where U_N is a $p \times 1$ random vector such that

$$EU_N = 0$$

$$E(U_N U_N^T) = V_{NN}.$$

Also, let $\bar{X}_N$ be an unbiased estimator of x, independent of U_N such that

$$E[(\bar{X}_N - x)(\bar{X}_N - x)^T] = Q_{NN}$$

and

$$E[(\bar{X}_N - x)U_N^T] = 0.$$

Then the best linear estimator for x is

$$\hat{X} = [Q_{NN}^{-1} + H_N^T V_{NN}^{-1} H_N]^{-1}[Q_{NN}^{-1}\bar{X}_N + H_N^T V_{NN}^{-1} Y_N] \qquad (3\text{-}36)$$

Proof: Let

$$Y = \begin{bmatrix} \bar{X}_N \\ Y_N \end{bmatrix}, \quad H = \begin{bmatrix} I \\ H_N \end{bmatrix}, \quad \text{and} \quad U = \begin{bmatrix} \bar{X}_N - x \\ U_N \end{bmatrix}.$$

Then

$$V = \begin{bmatrix} Q_{NN} & \phi \\ \phi & V_{NN} \end{bmatrix} \quad \text{and} \quad V^{-1} = \begin{bmatrix} Q_{NN}^{-1} & 0 \\ 0 & V_{NN}^{-1} \end{bmatrix}.$$

The formula (3-36) is the result of a direct substitution of these quantities into (3-33) and performing the proper manipulations.

Note that if one adds and subtracts $\bar{X}_N$ from the right side of (3-36), it yields a most useful *recursive* form:

$$\hat{X} = \bar{X}_N + [Q_{NN}^{-1} + H_N^T V_{NN}^{-1} H_N]^{-1} H_N^T V_{NN}^{-1}[Y_N - H_N \bar{X}_N].$$

For further development of linear estimation consult the References at the back of this book.

EXERCISES

1. Prove Corollary 3-1.2.

2. Prove Corollaries 3-2.1–3-2.4.

3. Show that $\hat{X}^{(1)}$ in Theorem 3-7.1 is equal to

$$\hat{X}^{(1)} = [H_1^T V^{-1} H_1 - H_1^T V^{-1} H_2 (H_2^T V^{-1} H_2)^{-1} H_2^T V^{-1} H_1]^{-1} H_1^T V^{-1} Y$$
$$- (H_1^T V^{-1} H_1)^{-1} H_1^T V^{-1} H_2 [H_2^T V^{-1} H_2$$
$$- H_2^T V^{-1} H_1 (H_1^T V^{-1} H_1)^{-1} H_1^T V^{-1} H_2]^{-1} H_2^T V^{-1} Y.$$

4. Given the linear model

$$Y = (1, h_1, h_2) \begin{pmatrix} x_1 \\ x_2 \\ x_3 \end{pmatrix} + U,$$

where $E(U) = \phi$, $E(UU^T) = \begin{bmatrix} 1 & 2 & 1 \\ 2 & 8 & 4 \\ 1 & 4 & 6 \end{bmatrix}$. Find estimates for x_1, x_2, x_3 by using the Gauss–Markov theorems for the following data:

y	1	5	0	4	4	-1
h_1	0	2	1	3	2	3
h_2	1	2	2	1	1	3

5. Let $Y_1, Y_2, \ldots, Y_N$ be random vectors. Let $\sum_{i=1}^{N} a_i Y_i$ be an estimator for the random vector X, where a_i, $i = 1, 2, \ldots, N$ are constants. Show that the constants a_i, $i = 1, 2, \ldots, N$ that minimize

$$E\left[\left(\sum_{i=1}^{N} a_i Y_i - X\right)\left(\sum_{i=1}^{N} a_i Y_i - X\right)^T\right]$$

are such that

$$E\left[\left(X - \sum_{i=1}^{N} a_i Y_i\right) Y_i\right] = 0, \quad i = 1, 2, \ldots, N.$$

6. Given the linear model

$$Y = x_1 + \sum_{i=1}^{3} h_i x_{i+1} + U,$$

where $(x_1, x_2, x_3, x_4)^T$ is a 4×1 parameter vector and

$$E(U) = \phi$$

$$E(UU^T) = \begin{bmatrix} 1 & 1 & 1 & 1 \\ 1 & 5 & 3 & 3 \\ 1 & 3 & 3 & 3 \\ 1 & 3 & 3 & 7 \end{bmatrix}.$$

Find estimates for x_1, x_2, x_3, and x_4 by using the Gauss–Markov theorem for the following data:

Y	1	3	0	4	1	2
h_1	1	1	1	1	1	1
h_2	2	0	1	0	2	3
h_3	4	0	2	0	4	6
h_4	1	1	1	2	2	2

CHAPTER 4

Best Linear Recursive Estimation

4-1 Introductory Remarks

In many problems of estimation, computational difficulties arise because of large amounts of data or because of having to perform on-line or real-time parameter estimations. Attempts to lighten this computational difficulty have led to recursive schemes [4, 13, 22, 25, 26] such that it is not necessary to store all the previous data but only previous estimates and current data. In this chapter we shall present a practical technique for obtaining a best linear unbiased estimator for the parameter vector in Equation (4-1).

Let Y_i be a $p \times 1$ random vector being observed at time t_i, $i = 1, 2, \ldots, N$. Let

$$Y_i = h_i x + U_i \qquad i = 1, 2, \ldots, N \qquad (4\text{-}1)$$

be a linear model relating the random vector, Y_i, of observations with the $n \times 1$ parameter vector x. The $p \times n$ matrix h_i, $i = 1, 2, \ldots, N$ is assumed nonrandom and known, while the $p \times 1$ error vector U_i is such that

$$EU_i = 0 \qquad \text{for every } i = 1, 2, \ldots, N \qquad (4\text{-}2)$$

and

$$E[U_i U_j^T] = V_{ij}, \qquad i, j = 1, 2, \ldots, N, \qquad (4\text{-}3)$$

where V_{ij} is a known $p \times p$ covariance matrix.

The model in (4-1) can be written in matrix notation in the following manner:

$$T_N = H_N x + E_N, \tag{4-4}$$

where

$$T_N = \begin{bmatrix} Y_1 \\ Y_2 \\ \cdot \\ \cdot \\ \cdot \\ Y_{N-1} \\ Y_N \end{bmatrix} = \begin{bmatrix} T_{N-1} \\ Y_N \end{bmatrix}$$

is an $Np \times n$ matrix of $p \times 1$ observation vectors Y_i, $i = 1, 2, \ldots, N$.

$$H_N = \begin{bmatrix} h_1 \\ h_2 \\ \cdot \\ \cdot \\ \cdot \\ h_{N-1} \\ h_N \end{bmatrix} = \begin{bmatrix} H_{N-1} \\ h_N \end{bmatrix}$$

is an $Np \times n$ matrix of $p \times n$ submatrices h_i, $i = 1, 2, \ldots, N$, and

$$E_N = \begin{bmatrix} U_1 \\ U_2 \\ \cdot \\ \cdot \\ \cdot \\ U_{N-1} \\ U_N \end{bmatrix} = \begin{bmatrix} E_{N-1} \\ U_N \end{bmatrix}$$

is an $Np \times 1$ matrix of $p \times 1$ random error vectors U_i, $i = 1, 2, \ldots, N$. The covariance of E_N is

$$E[E_N E_N^T] = V_N = \begin{bmatrix} V_{11} & \cdots & V_{1,N-1} & V_{1,N} \\ \cdot & & \cdot & \cdot \\ \cdot & & \cdot & \cdot \\ \cdot & & \cdot & \cdot \\ V_{N-1,1} & \cdots & V_{N-1,N-1} & V_{N-1,N} \\ V_{N,1} & \cdots & V_{N,N-1} & V_{N,N} \end{bmatrix} \tag{4-5}$$

Each $V_{ii}, i = 1, 2, \ldots, N$ will be assumed to be positive definite. The Gauss–Markov theorem states that the best linear unbiased estimator, say, $\hat{X}_N$, for the nonrandom parameter vector x is given by

$$\hat{X}_N = [H_N^T V_N^{-1} H_N]^{-1} H_N V_N^{-1} T_N. \qquad (4\text{-}6)$$

However, if the restriction that V_{ij} in (4-3) is such that

$$V_{ij} = V_{ij} \delta_{ij},$$

where

$$\delta_{ij} = 1 \quad \text{if } i = j$$
$$= 0 \quad \text{if } i \neq j$$

then V_n in (4-5) can be written as

$$V_N = \begin{bmatrix} V_{N-1} & 0 \\ 0 & V_{NN} \end{bmatrix} \qquad (4\text{-}7)$$

and (4-6) reduces to the recursive form

$$\hat{X}_N = \hat{X}_{N-1} + [H_{N-1}^T V_{N-1}^{-1} H_{N-1} + h_N^T V_{NN}^{-1} h_N]^{-1} h_N^T V_{NN}^{-1} [Y_N - h_N \hat{X}_{N-1}] \qquad (4\text{-}8)$$

and from Theorem 1-4.9:

$$[H_N^T V_N^{-1} H_N]^{-1} = [H_{N-1}^T V_{N-1}^{-1} H_{N-1} + h_N^T V_{NN}^{-1} h_N]^{-1}$$
$$= (H_{N-1}^T V_{N-1}^{-1} H_{N-1})^{-1} - (H_{N-1}^T V_{N-1}^{-1} H_{N-1})^{-1} h_N^T [V_{NN}^{-1}$$
$$+ h_N (H_{N-1}^T V_{N-1}^{-1} H_{N-1})^{-1} h_N^T]^{-1} h_N (H_{N-1}^T V_{N-1}^{-1} H_{N-1})^{-1}. \qquad (4\text{-}9)$$

The storage space required to compute (4-6) is approximately $[(N - 1)/2]p \cdot (p + 2n + 3)$ elements of past data plus present data, while the storage space required to compute (4-8) is $n + (p^2 + p)/2$ elements of past data plus present data. If N were large, then a considerable amount of storage space could be saved by using Equation (4-8). It is important to note that (4-9) allows one to compute the covariance matrix of $\hat{X}_N$ in (4-9) by inverting the $p \times p$ matrix:

$$V_{NN}^{-1} + h_N (H_{N-1}^T V_{N-1}^T H_{N-1})^{-1} h_N^T.$$

Another useful form of $\hat{X}_N$ when (4-7) is true is

$$\hat{X}_N = \left[\sum_{i=1}^{N} h_i^T V_{ii}^{-1} h_i\right]^{-1} \left[\sum_{i=1}^{N} h_i^T V_{ii}^{-1} Y_i\right]$$

$$= \left[\sum_{i=1}^{N-K} h_i^T V_{ii}^{-1} h_i + \sum_{i=N-K+1}^{N} h_i^T V_{ii}^{-1} h_i\right]^{-1} \left[\sum_{i=1}^{N-K} h_i^T V_{ii}^{-1} Y_i \right.$$

$$\left. + \sum_{i=N-K+1}^{N} h_i^T V_{ii}^{-1} Y_i\right]$$

$$= [P_{1,N-K}^{-1} + P_{N-K+1,N}^{-1}]^{-1} [P_{1,N-K}^{-1} \hat{X}_{N-K} + P_{N-K+1,N}^{-1} X_N^*],$$

where

$$P_{1,N-K} = \left[\sum_{i=1}^{N-K} h_i^T V_{ii}^{-1} h_i\right]^{-1}$$

and

$$P_{N-K+1,N} = \left[\sum_{i=1}^{N-K} h_i^T V_{ii}^{-1} h_i\right]^{-1}.$$

$P_{1,N-K}$ and $P_{N-K+1,N}$ are the covariance matrices of $\hat{X}_{N-K}$ and X_N^*, the best linear unbiased estimators for x, using the first $N - K$ observations and the least K observations, respectively.

4-2 A Useful Generalization

We shall consider in this chapter a more general case, in which V_{ij} may not be null for all $i \neq j$.

Suppose that

$$E(E_N E_N^T) = \{V_{ij}\}, \qquad i, j = 1, 2, \ldots, N$$

and let

$$V_N = \{V_{ij}\}$$

be the covariance matrix of E_N, where V_{ij} is a $p \times p$ matrix not necessarily null for $i \neq j$. Again the minimum variance linear unbiased estimator for x is given by

$$\hat{X}_N = (H_N^T V_N^{-1} H_N)^{-1} H_N^T V_N^{-1} T_N. \tag{4-10}$$

For (4-10) to be useful, V_N is assumed to be known, or equivalently, V_N^{-1} is

Sec. 4-2 A Useful Generalization

known. Since V_N^{-1} is positive definite, it can be written as a result of a Crout factorization as

$$V_N^{-1} = P_N P_N^T, \tag{4-11}$$

where P_N can be generated recursively with respect to the subscript N. We mean by the statement that P_N is generated recursively, that the $(N-1)p$th row of P_N does not depend on the Npth row of V_N for all N; that is, the elements of P_{N-1} do not depend on the Npth row of V_N. The matrix P_N is a lower triangular matrix with positive elements along the principal diagonal. Substituting the right side of (4-11) into (4-10) one can rewrite (4-10):

$$\hat{X} = (G_N^T G_N)^{-1} G_N^T z_N, \tag{4-12}$$

where

$$G_N = P_N^T H_N \tag{4-13}$$

$$Z_N = P_N^T T_N. \tag{4-14}$$

Consider the following partitioning of

$$P_N = \begin{bmatrix} P_{N-1} & 0 \\ p_N & p_{NN} \end{bmatrix}, \tag{4-15}$$

where the $(N-1)p \times (N-1)p$ matrix P_{N-1} is such that

$$V_{N-1}^{-1} = P_{N-1} P_{N-1}^T,$$

p_N is a $p \times (N-1)p$ matrix, and p_{NN} is a $p \times p$ lower triangular matrix.

It follows from (4-13) and (4-15) that

$$G_N = \begin{bmatrix} P_{N-1}^T & p_N^T \\ 0 & p_{NN}^T \end{bmatrix} \begin{bmatrix} H_{N-1} \\ h_N \end{bmatrix}$$

$$= \begin{bmatrix} G_{N-1} + p_N^T h_N \\ p_{NN}^T h_N \end{bmatrix}$$

and similarly from (4-14) and (4-14) that

$$Z_N = \begin{bmatrix} P_{N-1}^T T_{N-1} + p_N^T Y_N \\ p_{NN}^T Y_N \end{bmatrix}$$

$$= \begin{bmatrix} Z_{N-1} + p_N^T Y_N \\ p_{NN}^T Y_N \end{bmatrix}.$$

Note that the dimensions of G_N and Z_N are $Np \times n$ and $Np \times 1$, respectively. It follows that

$$G_N^T G_N = G_{N-1}^T G_{N-1} + G_{N-1}^T p_N^T h_N + h_N^T p_N G_{N-1} \\ + h_N^T p_N p_N^T h_N + h_N^T p_{NN} p_{NN}^T h_N \quad (4\text{-}16)$$

and

$$G_N^T Z_N = G_{N-1}^T Z_{N-1} + h_N^T p_N Z_{N-1} + G_{N-1}^T p_N^T Y_N \\ + h_N^T p_N p_N^T Y_N + h_N^T p_{NN} p_{NN}^T Y_N. \quad (4\text{-}17)$$

Both of these quantities cannot be generated recursively since

$$h_N^T p_N Z_{N-1}, \qquad G_{N-1}^T p_N^T Y_N \quad (4\text{-}18)$$

and

$$h_N^T p_N G_{N-1} \quad (4\text{-}19)$$

require every element of Z_{N-1} and G_{N-1} to compute these quantities.

Hence we can conclude that unless further assumptions can be made concerning the form of the covariance matrix V_N a recursive form of the estimator $\hat{X}_N$ cannot be formulated from the approach that we have taken.

A realistic assumption that yields a recursive estimator, yet requires one to retain for computing several latter elements of Z_{N-1} and G_{N-1}, is the assumption that

$$V_{ij} = 0$$

for all i and j such that $|i - j| > c$. For example, suppose that $c = 2$; then V_N is of the following form:

$$V_N = \begin{bmatrix} V_{11} & V_{12} & V_{13} & 0 & 0 & 0 & \cdots & 0 & 0 & 0 & 0 \\ V_{21} & V_{22} & V_{23} & V_{24} & 0 & 0 & \cdots & 0 & 0 & 0 & 0 \\ V_{31} & V_{32} & V_{33} & V_{34} & V_{35} & 0 & \cdots & 0 & 0 & 0 & 0 \\ 0 & V_{42} & V_{43} & V_{44} & V_{45} & V_{46} & \cdots & 0 & 0 & 0 & 0 \\ \vdots & \vdots & \vdots & \vdots & \vdots & \vdots & & \vdots & \vdots & \vdots & \vdots \\ 0 & 0 & 0 & 0 & 0 & 0 & \cdots & 0 & V_{N,N-2} & V_{N,N-1} & V_{NN} \end{bmatrix}$$

It can be shown that, if $V_{ij} = 0$ for $|i - j| > c$, $P_N = \{p_{ij}\}$ is such that

$$p_{ij} = 0,$$

Sec. 4-2 A Useful Generalization

and $|i - j| > c$. This is fortunate indeed since this implies that p_N can be written in the following form:

$$p_N = [\phi_N \bar{p}_N],$$

where ϕ is a $p \times Np - (cp + p)$ null matrix and $\bar{p}_N$ is a $p \times cp$ matrix. We can now write

$$G_N^T G_N = G_{N-1}^T G_{N-1} + \bar{G}_{N-1}^T \bar{p}_N^T h_N + h_N^T \bar{p}_N \bar{G}_{N-1}$$
$$+ h_N^T \bar{p}_N \bar{p}_N^T h_N + h_N^T p_{NN} p_{NN}^T h_N$$

and

$$G_N^T Z_N = (G_{N-1}^T G_{N-1})\hat{X}_{N-1} + h_N^T \bar{p}_N \bar{Z}_{N-1} + \bar{G}_{N-1}^T \bar{p}_N^T Y_N$$
$$+ h_N^T \bar{p}_N \bar{p}_N^T Y_N + h_N^T p_{NN} p_{NN}^T Y_N,$$

where

$$\hat{X}_{N-1} = (G_{N-1}^T G_{N-1})^{-1} G_{N-1}^T Z_{N-1}.$$

$\bar{G}_{N-1}$ and $\bar{Z}_{N-1}$ are the last cp rows of G_{N-1} and Z_{N-1}, respectively. Note that

$$\bar{Z}_N = S_p \begin{bmatrix} \bar{Z}_{N-1} + \bar{p}_N^T Y_N \\ p_{NN}^T Y_N \end{bmatrix}$$

and

$$\bar{G}_N = S_p \begin{bmatrix} \bar{G}_{N-1} + \bar{p}_N h_N \\ p_{NN}^T h_N \end{bmatrix},$$

where S_p is a $cp \times (c + 1)p$ "advancing" matrix defined by

$$S_p = \begin{bmatrix} 0 & I & 0 & \cdots & 0 \\ 0 & 0 & I & \cdots & 0 \\ \cdot & \cdot & \cdot & & \cdot \\ \cdot & \cdot & \cdot & & \cdot \\ \cdot & \cdot & \cdot & & \cdot \\ 0 & 0 & 0 & \cdots & I \end{bmatrix} = \{S_{ij}\},$$

where each partition S_{ij} is a $p \times p$ matrix $i = 1, \ldots, c$ and $j = 1, \ldots, c + 1$.

Finally,

$$\hat{X}_N = \left[G_{N-1}^T G_{N-1} + (\bar{G}_{N-1}^T h_N^T) K_N \begin{pmatrix} \bar{G}_{N-1} \\ h_N \end{pmatrix} \right]^{-1}$$

$$\left[(G_{N-1} G_{N-1}) \hat{X}_{N-1} + (G_{N-1}^T h_N^T) K_N \begin{pmatrix} \bar{Z}_{N-1} \\ Y_N \end{pmatrix} \right],$$

where

$$K_N = \begin{bmatrix} 0 & \bar{p}_N^T \\ \bar{p}_N & \bar{p}_N \bar{p}_N^T + p_{NN} p_{NN}^T \end{bmatrix} \quad (4\text{-}20)$$

and

$$G_{N-1}^T G_{N-1} + (\bar{G}_{N-1}^T h_N^T) K_N \begin{pmatrix} \bar{G}_{N-1} \\ h_N \end{pmatrix}$$

can be inverted recursively as a function of $(G_{N-1}^T G_{N-1})^{-1}$ by applying the inside-out rule. The storage space required to compute (4-20) is approximately $(n/2)(n + 3 + 2cp) + cp$ elements of past data plus present data.

4-3 An Example

To clarify the partitioning of the matrices in Section 4-2, consider the following example, where $p = 2$, $c = 2$, and $N = 4$. Let

$$P_4 = \begin{bmatrix} 1 & 0 & 0 & 0 & 0 & 0 & 0 & 0 \\ 1 & 1 & 0 & 0 & 0 & 0 & 0 & 0 \\ 1 & 1 & 1 & 0 & 0 & 0 & 0 & 0 \\ 1 & 1 & 1 & 1 & 0 & 0 & 0 & 0 \\ 1 & 1 & 1 & 1 & 1 & 0 & 0 & 0 \\ 1 & 1 & 1 & 1 & 1 & 1 & 0 & 0 \\ \hline 0 & 0 & 1 & 1 & 1 & 1 & 1 & 0 \\ 0 & 0 & 1 & 1 & 1 & 1 & 1 & 1 \end{bmatrix};$$

hence

$$p_3 = \begin{bmatrix} 1 & 0 & 0 & 0 & 0 & 0 \\ 1 & 1 & 0 & 0 & 0 & 0 \\ 1 & 1 & 1 & 0 & 0 & 0 \\ 1 & 1 & 1 & 1 & 0 & 0 \\ 1 & 1 & 1 & 1 & 1 & 0 \\ 1 & 1 & 1 & 1 & 1 & 1 \end{bmatrix};$$

$$p_4 = \begin{bmatrix} 0 & 0 & 1 & 1 & 1 & 1 \\ 0 & 0 & 1 & 1 & 1 & 1 \end{bmatrix};$$

$$p_{44} = \begin{bmatrix} 1 & 0 \\ 1 & 1 \end{bmatrix}; \quad \bar{p}_4 = \begin{bmatrix} 1 & 1 & 1 & 1 \\ 1 & 1 & 1 & 1 \end{bmatrix};$$

$$\Phi_4 = \begin{bmatrix} 0 & 0 \\ 0 & 0 \end{bmatrix};$$

$$V^{-1} = P_4 P_4^T = \begin{bmatrix} 1 & 1 & 1 & 1 & 1 & 1 & 0 & 0 \\ 1 & 2 & 2 & 2 & 2 & 2 & 0 & 0 \\ 1 & 2 & 3 & 3 & 3 & 3 & 1 & 1 \\ 1 & 2 & 3 & 4 & 4 & 4 & 2 & 2 \\ 1 & 2 & 3 & 4 & 5 & 5 & 3 & 3 \\ 1 & 2 & 3 & 4 & 5 & 6 & 4 & 4 \\ 0 & 0 & 1 & 2 & 3 & 4 & 5 & 5 \\ 0 & 0 & 1 & 2 & 3 & 4 & 5 & 6 \end{bmatrix}$$

and finally, K_4 in (4-20) is

$$\begin{bmatrix} 0 & 0 & 0 & 0 & 1 & 1 \\ 0 & 0 & 0 & 0 & 1 & 1 \\ 0 & 0 & 0 & 0 & 1 & 1 \\ 0 & 0 & 0 & 0 & 1 & 1 \\ \hline 1 & 1 & 1 & 1 & 5 & 5 \\ 1 & 1 & 1 & 1 & 5 & 6 \end{bmatrix}.$$

There are other forms of V_N that admit recursive estimators of x. The total family of V_N's for which various forms of recursive estimators exist has still essentially not been characterized.

EXERCISES

1. Prove the statement that if P is a lower-block triangular matrix, then P^{-1} is a lower-block triangular matrix.

2. Prove the statement that if V is positive definite and block diagonal, the V^{-1} is positive definite and block diagonal.

3. Let

$$y_i = \begin{bmatrix} i \\ i+1 \end{bmatrix} \quad x = \begin{bmatrix} x_1 \\ x_2 \\ x_3 \end{bmatrix} \quad h_i = \begin{bmatrix} 1 & 1 & 1 \\ 1 & 2 & 4 \end{bmatrix}$$

$$v_{ij} = \begin{bmatrix} 1 & 0 \\ 0 & 1 \end{bmatrix} \quad \text{for all } i$$

$$v_{ij} = \begin{bmatrix} \frac{1}{2} & \frac{1}{2} \\ \frac{1}{2} & \frac{1}{2} \end{bmatrix} \quad \text{for all } |i - j| = 1$$

$$= 0 \quad \text{for all } |i - j| > 1.$$

Obtain $\hat{X}_4$.

4. Let
$$V^{-1} = \begin{bmatrix} 1 & 1 & 1 \\ 1 & 5 & 5 \\ 1 & 5 & 14 \end{bmatrix}.$$

Use the Crout fact factorization to find P and P^T such that $V^{-1} = PP^T$.

5. Verify that
$$\bar{G}_N = S_p \begin{bmatrix} \bar{G}_{N-1} + \bar{p}_N h_N \\ p_{NN}^T h_N \end{bmatrix}$$

and

$$\bar{Z}_N = S_p \begin{bmatrix} \bar{Z}_{N-1} + \bar{p}_N Y_N \\ p_{NN}^T h_N \end{bmatrix},$$

where S_p has been previously defined.

6. Let $Y_i = h_i x + U_i$, where Y_i is a 2×1 observable random vector, h_i is a 2×2 regression matrix, x is 2×1 parameter vector, and U_i is a 2×1 error vector such that

$$E(U_i) = \phi$$

$$E(U_i U_i^T) = V_{ii}.$$

Let $i = 1, 2, 3, 4$ and the V's be equal to the example of this chapter. Let

$$h_i = \begin{bmatrix} 1 & i \\ 1 & i+1 \end{bmatrix}, \quad i = 1, 2, 3, 4$$

and

$$y_1 = \begin{bmatrix} 1 \\ 2 \end{bmatrix}, \quad y_2 = \begin{bmatrix} 0 \\ 1 \end{bmatrix}, \quad y_3 = \begin{bmatrix} 2 \\ 1 \end{bmatrix}, \quad \text{and } y_4 = \begin{bmatrix} 1 \\ 1 \end{bmatrix}.$$

Find $\hat{X}_4$ by the method of this chapter.

CHAPTER 5

Best Linear Unbiased Estimation Using Continuous Data

5-1 Introduction

In many applied problems data is obtained in a continuous form. That is, the $p \times 1$ data vector is composed of elements that are continuous functions of a real scalar variable t, usually denoting time. Also, in many of these applied problems, the parameters $x_1(t), \ldots, x_n(t)$ are assumed nonrandom and functions of the scalar real variable t. To facilitate the development to follow we give two definitions.

Definition 5-1.1. *The model*

$$Y_j(t) = \sum_{m=1}^{n} \int_{-\infty}^{\infty} h_{jm}(t-u) x_m(u) \, du + V_j(t)$$
(5-1)
$$j = 1, \ldots, p \quad -\infty < t < \infty,$$

where $\{h_{jk}(t)\}$ *is a* $p \times n$ *matrix of known nonrandom functions and* $U_j(t)$ *is a jointly weakly stationary stochastic noise process; that is,*

$$E[V_j(t)] = 0 \quad \text{for all } j \text{ and } t$$

and

$$E[V_i(t)V_j(t')] = R_{ij}(t - t').$$
(5-2)

The model described in Definition 5-1.1 is called the multivariate general linear model for continuous data.

As in the discrete case we must define what is meant by a linear estimator of the parameter function vector, $x(t) = [x_1(t), \ldots, x_n(t)]^T$, in the general linear model defined by (5-1). A basic regularity condition tacitly assumed here is that the Fourier transforms of $h_{jk}(t)$ and $x_j(t)$ are assumed to exist for all j and k.

Definition 5-1.2. *The class of estimators*

$$L(\hat{X}) = \left\{ \hat{X}_j(t): \hat{X}_j(t) = \sum_{k=1}^{n} \int_{-\infty}^{\infty} b_{jk}(s) Y_k(t-s)\, ds \right\}, \quad (5\text{-}3)$$

where $b_{jk}(s)$ is a function of real scalar variable s and possesses a Fourier transform, is said to be the class of linear estimators for $x_j(t)$. The matrix $\{b_{jk}(t)\}$ is an n × p matrix called by many engineering scientists a matrix of "filter functions."

5-2 A Best Linear Unbiased Estimator

Since $x(t) = [x_1(t), x_2(t), \ldots, x_n(t)]^T$ is a vector it is convenient to select a vector $a(t) = [a_1(t), \ldots, a_n(t)]^T$ such that if

$$Z(t) = \sum_{j=1}^{n} \int_{-\infty}^{\infty} a_j(t) \hat{X}_j(t-u)\, du$$

then

$$E\{Z(t) - E[Z(t)]\}^2$$

is minimal.

Note first that

$$Z(t) - E[Z(t)] = \sum_{j,k} \int_{-\infty}^{\infty} \int_{-\infty}^{\infty} a_j(u) b_{jk}(s) V_k(t-u-s)\, du\, ds, \quad (5\text{-}4)$$

from which by squaring and taking expectation one can determine the variance of $Z(t)$. By requiring that

$$\hat{X}_j(t) = \sum_{k=1}^{n} \int_{-\infty}^{\infty} b_{jk}(s) Y_k(t-s)\, ds \quad (5\text{-}5)$$

be an unbiased estimator of $x_j(t)$, substituting (5-1) into (5-5), and imposing the unbiased constraint it follows that

$$x_j(t) = \sum_{k} \sum_{m} \int_{-\infty}^{\infty} \int_{-\infty}^{\infty} b_{jk}(s) h_{km}(t-s-u) x_m(u)\, du\, ds, \quad (5\text{-}6)$$

Sec. 5-2 A Best Linear Unbiased Estimator

which implies that

$$\hat{X}_j(t) = x_j(t) + \sum_{k=1}^{n} \int_{-\infty}^{\infty} b_{jk}(s) V_k(t-s) \, ds.$$

Taking the Fourier transforms of Equation (5-6) one obtains the unbiased condition in terms of Fourier transforms; that is,

$$B_{jk}(\omega) H_{km}(\omega) X_m(\omega)$$

$$= \int_{-\infty}^{\infty} \int_{-\infty}^{\infty} b_{jk}(s) \left[\int_{-\infty}^{\infty} h_{km}(t-s-u) x_m(u) \, du \right] ds \, e^{-i\omega t} \, dt$$

or

$$\frac{1}{2\pi} \int_{-\infty}^{\infty} B_{jk}(u) H_{km}(\omega) X_m(\omega) e^{i\omega t} \, d\omega$$

$$= \int_{-\infty}^{\infty} b_{jk}(s) \left[\int_{-\infty}^{\infty} h_{km}(t-s-u) x_m(u) \, du \right] ds,$$

which implies that

$$x_j(t) = \sum_m \sum_k \int_{-\infty}^{\infty} B_{jk}(\omega) H_{km}(\omega) X_m(\omega) e^{i\omega t} \frac{d\omega}{2\pi}. \tag{5-7}$$

Thus Equation (5-7) implies that unbiasedness will hold only if

$$\sum_{k=1}^{n} B_{jk}(\omega) H_{km}(\omega) = \delta_{jm}, \tag{5-8}$$

where δ_{jm} is the Kronecker delta operator.

Equation (5-8) can be written in matrix notation as

$$B(\omega) H(\omega) = I, \tag{5-9}$$

where I is an $n \times n$ identity matrix. The independent variate ω can be thought of a frequence and (5-9) can be thought as an identity at each frequency ω.

From (5-4) it follows that

$$E\{[Z(t) - EZ(t)][Z(t') - EZ(t')]\}$$

$$= \sum_{j,k} \sum_{j',k'} \int_{-\infty}^{\infty} \int_{-\infty}^{\infty} \int_{-\infty}^{\infty} \int_{-\infty}^{\infty} a_j(u) b_{jk}(s) R_{kk'}(\gamma - u - s + u' + s')$$

$$\times b_{j'k'}(s') a_{j'}(u') \, du \, ds \, du' \, ds'$$

$$= \sum_{j,k} \sum_{j',k'} a_j(\gamma) * b_{jk}(\gamma) * R_{kk'}(\gamma) * b_{j',k'(-\gamma)} * a_{j'}(-\gamma), \tag{5-10}$$

where $\gamma = t - t'$ and $*$ indicates convolution. Thus by taking the Fourier transform and then the inverse Fourier transform of (5-10) we can write

$$E\{[Z(t) - EZ(t)][Z(t') - EZ(t')]^T\}$$

$$= \sum_{j,k} \sum_{j',k'} \int_{-\infty}^{\infty} A_j(\omega) B_{jk}(\omega) S_{kk'}(\omega) B_{j'k'}(-\omega) A_{j'}(-\omega) e^{i\omega\gamma} \frac{d\omega}{2\pi}$$

$$= \int_{-\infty}^{\infty} ABSB^* A^* e^{i\omega\gamma} \frac{d\omega}{2\pi}, \qquad (5\text{-}11)$$

where $A = [A_1(\omega), A_2(\omega), \ldots, A_n(\omega)]$, $B = B_{jk}(\omega)$ is an $n \times p$ matrix, and $S = [S_{kk'}(\omega)]$ is a $p \times p$ spectra matrix. By letting $\gamma = 0$ we obtain the variance of $Z(t)$, that is,

$$E[Z(t) - EZ(t)]^2 = \int_{-\infty}^{\infty} ABSB^* A^* \frac{d\omega}{2\pi} \qquad (5\text{-}12)$$

To determine matrix $B(\omega)$ such that the variance of $Z(t)$ is minimal, Theorem 5-1.1 is helpful.

THEOREM 5-2.1. *If* C *is a* p $\times$ p *positive definite matrix,* H *is a* p $\times$ n *matrix* n $\leq$ p *of rank* n, *and* α *is any* 1 $\times$ p *complex vector, then*

$$\alpha^* C^{-1} \alpha \geq \alpha^* H (H^* C H)^{-1} H^* \alpha. \qquad (5\text{-}13)$$

Proof: Let U be a unitary matrix; that is,

$$U^* U = UU^* = I$$

such that

$$UC^{-1}U^* = \lambda^{-1} I,$$

where λ is a scalar. Then it follows that

$$UCU^* = \lambda I.$$

Let

$$\alpha^* U^* = \beta^*$$

and

$$Y = UH;$$

then

$$H = U^* Y.$$

Sec. 5-2 A Best Linear Unbiased Estimator

It now follows that

$$\alpha^* U^* U C^{-1} U^* U \alpha - \alpha^* U^* Y (Y^* U C U^* Y)^{-1} Y^* U \alpha$$
$$= \beta^* \lambda^{-1} I \beta - \beta^* Y (Y^* \lambda I Y)^{-1} Y^* \beta$$
$$= \lambda^{-1} \beta^* [I - Y(Y^* Y)^{-1} Y^*] \beta.$$

We note that

$$I - Y(Y^* Y)^{-1} Y^*$$

is symmetric idempotent, hence positive semidefinite, and (5-13) is true.

Select $\alpha = AB$, and $S = C^{-1}$; then

$$\text{Var } Z(t) \geq \int_{-\infty}^{\infty} ABH(H^* S^{-1} H)^{-1} H^* B^* A^* \frac{d\omega}{2\pi}$$
$$= \int_{-\infty}^{\infty} A(H^* S^{-1} H)^{-1} A^*(\omega) \frac{d\omega}{2\pi} \quad (5\text{-}14)$$

when the unbiased condition that $BH = I$ in (5-9) is imposed. Comparing (5-13) and (5-14) it follows that equality is obtained when

$$B(\omega) = (H^* S^{-1} H)^{-1} H^* S^{-1}.$$

The covariance matrix of $\hat{x}(t)$ follows in an analysis similar to the above as

$$\{\text{Cov } \hat{X}_i(t), \hat{X}_j(t)\} = \int_{-\infty}^{\infty} BSB^* \frac{d\omega}{2\pi} = \int_{-\infty}^{\infty} (H^* S^{-1} H)^{-1} \frac{d\omega}{2\pi}.$$

Consider the simple model

$$Y_j(t) = s(t) + V_j(t) \quad j = 1, 2, \ldots, p,$$

where $s(t)$ is a nonrandom continuous signal and $V_j(t)$ is a jointly weakly stationary zero-mean multivariate noise process with correlation function $R(t - s)$. We can rewrite the above equation as

$$Y_j(t) = \int_{-\infty}^{\infty} \delta(t - u) s(u) \, du + V_j(t) \quad i = 1, 2, \ldots, p,$$

which implies $h_{j1}(t) = \delta(t)$ and $x_1(t) = s(t)$. Thus

$$\int_{-\infty}^{\infty} \delta(t) e^{i\omega t} \, dt = 1 = H_{j1}(\omega),$$

which implies

$$H = \begin{pmatrix} 1 \\ 1 \\ \vdots \\ 1 \end{pmatrix},$$

a $p \times 1$ matrix. Hence one can now easily compute Var $[\hat{S}(t)]$.

5-3 A Useful Formulation When X(T) Is Random

Consider the following continuous model,

$$Y(t) = H(t)X(T) + V(t), \tag{5-15}$$

where $Y(t)$ is an observable $p \times 1$ random vector, t is a real variable such that $t_0 \leq t \leq T$, $H(t)$ is a $p \times n$ matrix whose elements are real functions of the variable t, and $X(T)$ is a $n \times 1$ random vector, which we wish to estimate, given that

$$\begin{aligned} E[X(T)] &= u \\ E\{[X(T) - u][X(T) - u]^T\} &= W. \end{aligned} \tag{5-16}$$

$V(t)$ is a $p \times 1$ random vector such that

$$E[V(t)] = 0 \quad \text{for } t_0 \leq t \leq T$$

$$E[V(t)V^T(s)] = R(t)\delta(t, s),$$

where $\delta(t, s)$ is the Dirac delta function, and

$$E[X(T)V^T(t)] = 0 \quad \text{for } t_0 \leq t \leq T.$$

Lemmas 5-3.1 and 5-3.2 will be developed in order to enhance the reading to follow.

LEMMA 5-3.1. *The covariance of* X(T) *and* Y(t) *for* $t_0 \leq t \leq T$ *is given by* $C[X(T), Y(t)] = WH^T(t)$.

Proof:

$$C[X(T), Y(t)] = E[X(T) - u][Y(t) - H(t)u]^T$$

$$= E[X(T) - u][H(t)X(T) + V(t) - H(t)u]^T$$

$$= E[X(T) - u][X(T) - u]^T H^T(t) + E[X(T) - u]V^T(t)$$

$$= WH^T(t) \quad \text{for } t_0 \leq t \leq T.$$

LEMMA 5-3.2. $C[Y(T), Y^T(s)] = R(t)\delta(t, s) + H(t)WH^T(s)$.

Proof:

$$E[Y(T) - H(t)u][Y(s) - H(s)u]^T$$

$$= E[H(t)X(T) + V(t) - H(t)u][H(s)X(T) + V(s) - H(s)u]^T$$

$$= E\{H(t)[X(T) - u] + V(t)\}\{H(s)[X(T) - u] + V(t)\}^T$$

$$= H(t)WH^T(s) + R(t)\delta(t, s).$$

One can define a linear estimator of a linear combination of the elements of $X(T)$. Say,

$$D = p^T X(T)$$

by

$$D^* = a + \int_{t_0}^{T} n^T(t) Y(T) \, dt, \tag{5-17}$$

where p is a known $n \times 1$ vector of constants, a is a scalar constant, and $n(t)$ is an arbitrary $n \times 1$ vector function (at least piecewise continuous) of the real variable t. Also, we define a linear estimator for $p^T X(T)$ as "best" if

$$E[p^T X(T) - D^*] = 0 \tag{5-18}$$

and

$$E[p^T X(T) - D^*]^2 \tag{5-19}$$

is minimal.

The problem is to select an a and a vector-valued function $n(t)$ so that (5-18) is satisfied and (5-19) is minimized. We summarize the desired results in Theorem 5-3.1.

THEOREM 5-3.1. *If* $EX(T)$ *is unknown, then the best linear estimator for* $p^T X(T)$ *is*

$$D^* = p^T M^{-1}(t_0, T) \int_{t_0}^{T} H^T(t) R^{-1}(t) Y(t) \, dt, \tag{5-20}$$

where

$$M(t_0, T) = \int_{t_0}^{T} H^T(t)R^{-1}(t)H(t)\, dt, \qquad (5\text{-}21)$$

provided that the expectation operator and the integration operator commute.

Proof: Note that (5-20) implies that the scalar a in (5-17) is zero and

$$n(t) = R^{-1}(t)H(t)M^{-1}(t_0, T)p.$$

Consider the conditional expectation with respect to Y given X,

$$E_{Y|X}[D^*] = E_{Y|X}\left[p^T M^{-1}(t_0, T) \int_{t_0}^{T} H^T(t)R^{-1}(t)Y(t)\, dt \right]$$

$$= p^T M^{-1}(t_0, T) \int_{t_0}^{T} H^T(t)R^{-1}(t)H(t)\, dt\, X(T)$$

$$= p^T X(T),$$

which implies

$$E(D^* - p^T X(T)) = 0$$

and (5-18) holds. Let

$$D = a + \int_{t_0}^{T} n_1^T(t)Y(t)\, dt$$

be another linear estimator for $p^T X(T)$ for which (5-18) holds; then

$$E_{Y|X}[D] = a + \int_{t_0}^{T} n_1^T(t)H(t)\, dt\, X(T).$$

Condition (5-18) requires that

$$p^T X(T) = a + \int_{t_0}^{T} n_1^T(t)H(t)\, dt\, X(T)$$

$$0 = \left[p^T - \int_{t_0}^{T} n_1^T(t)H(t)\, dt \right] X(T) - a \qquad (5\text{-}22)$$

for every $X(T)$. The linearity of (5-22) implies that

$$p^T - \int_{t_0}^{T} n_1^T(t)H(t)\, dt = 0$$

$$a = 0.$$

Sec. 5-3 A Useful Formulation When X(T) Is Random

We note that

$$\int_{t_0}^{T} n_1^T(t) H(t)\, dt = p^T.$$

To minimize (5-19) consider

$$E[D - p^T X(T)]^2 = E\left[\int_{t_0}^{T} n_1^T(t) Y(t)\, dt - p^T X(T)\right]^2$$

$$= E\left[\int_{t_0}^{T} n^T(t) + r_n^T(t) Y(t)\, dt - p^T X(T)\right]^2$$

$$= E\left[D^* - p^T X(T) + \int_{t_0}^{T} r_n^T(t) Y(t)\, dt\right]^2$$

$$= E[D^* - p^T(T)]^2 + \int_{t_0}^{T} r_n^T(t) R(t) r_N(t)\, dt$$

$$- 2E\left[\int_{t_0}^{T} r_n^T(t) Y(t)\, dt\, X^T(T) p\right] + 2E\left[\int_{t_0}^{T} r_n^T(t) Y(t)\, dt\, D^*\right].$$

It follows that

$$E\left[\int_{t_0}^{T} r_n^T(t) Y(t)\, dt\, X^T(T) p\right] = \int_{t_0}^{T} r_n^T(t) H(t)(W + UU^T) p\, dt$$

by Lemma 5-3.1, and since

$$\int_{t_0}^{T} r_n^T(t) H(t)\, dt = \int_{t_0}^{T} [n_1^T(t) H(t) - n^T(t) H(t)]\, dt = p^T - p^T = 0,$$

then

$$E\left[\int_{t_0}^{T} r_n^T(t) Y(t)\, dt\, X^T(T) p\right] = 0.$$

Also, it follows that

$$E\left[\int_{t_0}^{T} r_n^T(t) Y(t)\, dt\, D^*\right] = E\left[\int_{t_0}^{T}\int_{t_0}^{T} r_n^T(t) Y(t) Y^T(s) n(s)\, dt\, ds\right]$$

$$= \int_{t_0}^{T}\int_{t_0}^{T} r_n^T(t)[H(t) W H^T(s) + R(t)\delta(t,s) + H(t) UU^T H^T(s)] n(s)\, dt\, ds.$$

by Lemma 5-3.2. But since

$$\int_{t_0}^{T} r_n^T(t) H(t)\, dt = 0,$$

then we have

$$\int_0^T r_n^T(s)R(s)n(s)\,ds = \int_{t_0}^T r_n^T(s)R(s)R^{-1}(s)H(s)M^{-1}(t_0,T)p\,ds$$

$$\int_{t_0}^T r_n^T(s)H(s)\,ds\,M^{-1}(t_0,T)p = 0.$$

Hence

$$E\left[\int_{t_0}^T r_n^T(t)Y(t)\,dt\,D^*\right] = 0.$$

Therefore,

$$E[D - p^T X(T)]^2 = E[D^* - p^T X(T)]^2 + \int_{t_0}^T r_n^T(t)R(t)r_n(t)\,dt,$$

which implies

$$E[D - p^T X(T)]^2 \geq E[D^* - p^T X(T)]^2$$

since $R(t)$ is positive definite and implies that

$$\int_{t_0}^T r_n^T(t)R(t)r_n(t)\,dt \geq 0.$$

Hence we conclude that D^* is best in the sense described above.

Since we have a best estimate for $p^T X(T)$, then by choosing $p = e_i$, the unit vector with 1 in the ith position, we see that the best estimate for $X(T)$ is

$$M(t_0,T)^{-1}\int_{t_0}^T H^T(t)R^{-1}(t)Y(t)\,dt, \qquad (5\text{-}23)$$

and the covariance matrix is simply $M(t_0,T)^{-1}$.

5-4 Relations Between Discrete and Continuous Models

Consider again the continuous linear model (5-15). As we have seen, the best estimator for the case where the mean was unknown was given by (5-23). The continuous estimator, heretofore, however was not well known and thus one was forced to model this model by a discrete model of the form

$$Y = HX(T) + V, \qquad (5\text{-}24)$$

Sec. 5-4 Relations Between Discrete and Continuous Models

where Y is an $Np \times 1$ vector, H an $Np \times n$ matrix, $X(T)$ and $n \times 1$ random vector such that

$$E[X(T)] = U,$$

an unknown quantity, and

$$E[X(T) - U][X(T) - U]^T = W,$$

a known $n \times n$ covariance matrix. The $Np \times 1$ vector V is a random vector such that

$$E[V] = \phi \quad \text{and} \quad E[VV^T] = R.$$

Once the conversion was made then an estimate was made for $X(T)$ by the estimator:

$$X^*(T) = (H^T R^{-1} H)^{-1} H^T R^{-1} Y. \tag{5-25}$$

The conversion to the discrete case was accomplished by taking N values of the variable t, say, $t_1, t_2, \ldots, t_N$; then

$$Y = \begin{bmatrix} Y(t_1) \\ Y(t_2) \\ \vdots \\ Y(t_N) \end{bmatrix}; \quad H = \begin{bmatrix} H(t_1) \\ H(t_2) \\ \vdots \\ H(t_N) \end{bmatrix}; \quad V = \begin{bmatrix} V(t_1) \\ V(t_2) \\ \vdots \\ V(t_N) \end{bmatrix}.$$

If follows that

$$R = \begin{bmatrix} R(t_1) & \phi & \cdots & \phi \\ \phi & R(t_2) & \cdots & \phi \\ \vdots & \vdots & & \vdots \\ \phi & \phi & \cdots & R(t_N) \end{bmatrix}.$$

Thus from (5-25) we can write

$$X^*(T) = \left[\sum_{i=1}^{N} H^T(t_i) R^{-1}(t_i) H(t_i) \right]^{-1} \left[\sum_{i=1}^{N} H^T(t_i) R^{-1}(t_i) Y(t_i) \right] \tag{5-26}$$

Consider now the estimator of $X(T)$ of the form of (5-23), where numerical integration must be used. To compute the integrals we take points $t_1, t_2, \ldots, t_N$ at equal intervals, say,

$$t_{i+1} - t_i = D;$$

then our numerical approximation for (5-23) becomes

$$\left[\sum_{i=1}^{N} H^T(t_i)R^{-1}(t_i)H(t_i)D\right]^{-1}\left[\sum_{i=1}^{N} H^T(t_i)R^{-1}(t_i)Y(t_i)\right]$$

or

$$\left[\sum_{i=1}^{N} H^T(t_i)R^{-1}(t_i)H(t_i)\right]^{-1}\left[\sum_{i=1}^{N} H^T(t_i)R^{-1}(t_i)Y(t_i)\right],$$

which is the same as Equation (5-26), which was the estimate when the continuous model was converted to the discrete model. Thus, if we numerically integrate the new estimator (5-23) or convert the continuous model to a discrete model with equal intervals between our points, we obtain the same results.

5-5 An Example

Consider the model

$$Y(t) = H(t)X(t) + V(t),$$

where $Y(t)$, $H(t)$, and $V(t)$ are defined as for model (5-15), but where $X(t)$ is a time series of a real variable t for

$$t_0 \leq t \leq T.$$

In some engineering applications $X(t)$ is characterized by the following differential equations:

$$\frac{dX(t)}{dt} = F(t)X(t) + G(t)N(t). \quad (5\text{-}27)$$

If continuity exists over an interval $[t_0, T]$, the solution to (5-27) is

$$X(t) = J(T, t)X(T) - J(T, t)\int_{t_0}^{T} J^{-1}(t, s)G(s)N(s)\,ds, \quad (5\text{-}28)$$

where $J(T, t)$ is a nonsingular matrix dependent on t. We may substitute (5-28) into the linear model to obtain

$$y(t) = H(t)J(T, t)X(T) + V(t) - J(T, t)\int_{t_0}^{T} J^{-1}(T, s)G(s)N(s)\,ds.$$

If $EX(T)$ is unknown, $N(s)$ is zero for all s, and the regular conditions of our theorem apply, then we have a best estimate for $X(T)$.

$$X^*(T) = M^{-1}(t_0, T) \int_{t_0}^T J^T(T, t) H^T(t) R^{-1}(t) Y(t)\, dt,$$

where

$$M(t_0, T) = \int_{t_0}^T J^T(T, t) H^T(t) R^{-1}(t) H(t) J(T, t)\, dt.$$

If $\mathring{N}(t)$ is not identically zero for all t, the dynamic model implies correlated $V(t)$ and our model here does not apply.

For suggested reading list, see References [15]–[18], [28], [32], [60], [78], [172], [201], and [203].

EXERCISES

1. Verify that

$$\{\text{Cov}\,[\hat{X}_i(t), \hat{X}_j(t)]\} = \int_{-\infty}^{\infty} (H^* \textstyle\sum^{-1} H)^{-1}\, \frac{d\omega}{2\pi}.$$

2. Given the model

$$Y_j = X_1(t) + X_2(t - T_j) + V_j(t) \qquad j = 1, 2, \ldots, p,$$

where $[V_j(t)]$ is a jointly weakly stationary process. Find h_{j1}, h_{j2}, and $\text{Cov}\,[\hat{X}_1(t), \hat{X}_2(t)]$.

3. Suppose the assumption

$$E[V(t)V^T(s)] = R(t)\delta(t, s)$$

in the linear model defined by (5-15) is changed to

$$E[V(t)V^T(s)] = R_V(\gamma),$$

where $\gamma = |s - t|$, $-\infty < \gamma < \infty$. Define the spectra of $V(t)$ to be

$$P_V(\omega) = \int_{-\infty}^{\infty} R_V(\gamma) e^{-i\gamma\omega}\, d\gamma$$

a. Show that if

$$N(t) = \int_{-\infty}^{\infty} g(u) V(t - u)\, du,$$

then

$$P_Y(\omega) = |G(\omega)|^2 P_V(\omega),$$

where $G(\omega)$ is the Fourier transform of $g(t)$ provided $P_V(\omega)$ exists.

b. If $N(t)$ is uncorrelated, that is,

$$E[N(t)] = 0$$

and

$$E[N(t)N^T(s)] = R_N(t)\delta(t-s),$$

then show that this requires $P_N(\omega) = c$, where c is constant for all ω. Further, show that this restriction requires

$$1 = |G(\omega)|^2 P_V(\omega).$$

What effect will this have on $P_V(\omega)$ and $G(\omega)$?

c. What is the autocorrelation function of $N(t)$?; that is, find $R_N(\gamma)$. How are $N(t)$ and $N(t+\gamma)$ correlated?

d. If $P_N(\omega)$ is such that $R_N(t)$ yields an arbitrarily small correlation for $\gamma \geq t_0$, where t_0 is an arbitrary positive number, then show that such a $P_N(\omega)$ gives

$$g(t) = \int_{-\infty}^{\infty} P_N^{1/2}(\omega) P_V^{-1/2}(\omega)] e^{i\omega t} \, d\omega,$$

if $P_V(\omega) \neq 0$.

4. Discuss how the following choice of $P_N(\omega)$ will satisfy Exercise 3d; that is, let

$$P_N(\omega) = 1, \quad -a \leq \omega \leq a$$

$$P_N(\omega) = 0 \quad \text{elsewhere.}$$

Then show that

$$R_N(\gamma) = \frac{\sin(a\gamma)}{\pi\gamma}$$

and

$$R_N(0) = \frac{a}{\pi}.$$

CHAPTER 6

On Linear Estimation with Constraints

6-1 Introductory Remarks

In this chapter we shall develop the best linear estimators of x in the linear model

$$Y = Hx + Z \tag{6-1}$$

under a set of constraints imposed on the unknown parameter vector x. We let H be an $m \times n$ full-rank matrix, Y an $m \times 1$ observable random vector, and Z an $m \times 1$ error vector such that $E(Z) = 0$ and $E(ZZ^T) = V$.

Three cases will be considered and are listed as follows:

Case 1. Estimate x given that the model (6-1) holds and further that the set of linear constraints on x are known to be such that

$$Ax = t, \tag{6-2}$$

where A and t are known $p \times n$ and $p \times 1$ matrices, respectively, such that

$$r(A) = p < n. \tag{6-3}$$

Case 2. Estimate x given that the model (6-1) holds and further that the set of linear constraints holds with an additive random component U

relating A, x, t, and U as

$$Sx + U = T,$$

where

$$E[U] = 0$$
$$E[UU^T] = P$$ (6-4)

with P known.

Case 3. Estimate x, where it is known that the following inequalities are true:

$$l \leq Ax \leq u,$$ (6-5)

where l and u are known nonrandom $n \times 1$ vectors and A is a known $p \times n$ matrix of any rank.

6-2 Estimation with NonRandom Linear Constraints

We consider Case 1 in this section. The result is summarized in Theorem 6-2.1.

THEOREM 6-2.1. *The minimum variance linear unbiased estimator for* x *in* (6-1) *given* (6-2) *is*

$$\tilde{X} = \hat{X} + (H^T V^{-1} H)^{-1} A^T [A(H^T V^{-1} H)^{-1} A^T]^{-1} (t - A\hat{X}),$$ (6-6)

where

$$\hat{X} = (H^T V^{-1} H)^{-1} H^T V^{-1} Y$$

Proof: It is well known that if x is unconstrained the, "payoff" function

$$Q = (y - Hx)^T V^{-1} (y - Hx)$$ (6-7)

when minimized will yield the minimum variance linear unbiased estimate for x. This estimate can be found as follows:

$$\frac{\partial Q}{\partial x} = -2H^T V^{-1}(y - Hx) = \Phi,$$

which implies

$$\hat{X} = (H^T V^{-1} H)^{-1} H^T V^{-1} Y.$$

To incorporate the constraints (6-2) we consider the Lagrangian equation

Sec. 6-2 NonRandom Linear Constraints

$$Q' = (y - Hx)^T V^{-1}(y - Hx) + 2\lambda^T(t - Ax),$$

where λ is an $m \times 1$ vector of Lagrangian multipliers. Then

$$\frac{\partial Q'}{\partial x} = -2H^T V^{-1}(y - Hx) - 2A^T \lambda = \phi$$

$$\frac{\partial Q'}{\partial \lambda} = t - Ax = \phi$$

(6-8)

gives necessary conditions for Q' to be a minimum. The equation (6-8) implies that

$$A^T \lambda = (H^T V^{-1} H)x - H^T V^{-1} y, \qquad (6-9)$$

which implies that the estimate will satisfy

$$(H^T V^{-1} H)x = H^T V^{-1} y + A^T \lambda,$$

or

$$\tilde{X} = \hat{X} + (H^T V^{-1} H)^{-1} A^T \lambda. \qquad (6-10)$$

To obtain λ multiply both sides of (6-9) by $A(H^T V^{-1} H)^{-1}$; that is, $A(H^T V^{-1} H)^{-1} A^T \lambda = Ax - A(H^T V^{-1} H)^{-1} H^T V^{-1} y = t - A(H^T V^{-1} H)^{-1} H^T V^{-1} y$. Note that the rank of A is such that $[A(H^T V^{-1} H)^{-1} A^T]$ is nonsingular. Hence

$$\lambda = [A(H^T V^{-1} H)^{-1} A^T]^{-1}[t - A\hat{X}]. \qquad (6-11)$$

Substituting (6-11) into (6-10) one obtains (6-6) and Theorem 6-2.1 is proved.

It is important to note that

$$t - A\tilde{X} = t - A[\hat{X} + (H^T V^{-1} H)^{-1} A^T \lambda]$$

$$= t - A\hat{X} - A(H^T V^{-1} H)^{-1} A^T \lambda$$

$$= t - A\hat{X} - (t - A\hat{X})$$

$$= \phi$$

as was required. Also note that $E(\tilde{X}) = x$ and the covariance matrix is obtained as follows:

$$\text{Cov}(\tilde{X}) = \text{Cov}\{\hat{X} + (H^T V^{-1} H)^{-1} A^T [A(H^T V^{-1} H)^{-1} A^T]^{-1}(t - A\hat{X})\}$$

$$= \text{Cov}\{(I - (H^T V^{-1} H)^{-1} A^T [A(H^T V^{-1} H)^{-1} A^T] A^{-1}) \hat{X}\}$$

$$= B(H^T V^{-1} H^T)^{-1} B^T,$$

where

$$B = I - (H^T V^{-1} H)^{-1} A^T [A(H^T V^{-1} H)^{-1} A^T]^{-1} A.$$

On reducing Cov $(\tilde{X})$ to a simpler form we find that

$$\text{Cov}(\tilde{X}) = (H^T V^{-1} H)^{-1} - (H^T V^{-1} H)^{-1} A^T [A(H^T V^{-1} H) A^T]^{-1} A(H^T V^{-1} H)^{-1}.$$

We note that additional information, that is, $Ax = t$, leads to a reduction in the variance of the estimator of x.

6-3 Linear Constraints with Additive Random Components

Suppose in addition to the linear model defined by (6-1), we have the conditions defined by (6-4) imposed on the unknown parameter vector x. Also, let the elements of U be stochastically independent of the elements of Z in (6-1); that is, $E[UZ^T] = \phi$. We further assume that S is a known $t \times n$ matrix, t is a known $t \times 1$ vector, and U is a $t \times 1$ random vector such that $E(UU^T) = P$ is a known positive definite matrix.

One can combine (6-1) and (6-4) to obtain a new linear model

$$\binom{Y}{T} = \binom{H}{S} x + \binom{Z}{U}$$

or

$$M = Gx + C, \tag{6-12}$$

where

$$M = \binom{Y}{T}, \quad G = \binom{H}{S}, \quad \text{and } C = \binom{Z}{U}.$$

We observe that

$$E(C) = \phi$$

$$E(CC^T) = K = \begin{bmatrix} V & \phi \\ \phi & P \end{bmatrix}$$

The Gauss–Markov estimator of x is $\tilde{X} = (G^T K^{-1} G)^{-1} G^T K^{-1} M$, which yields

$$\tilde{X} = [H^T V^{-1} H + S^T P^{-1} S]^{-1} [H^T V^{-1} Y + S^T P^{-1} T]$$
$$= \hat{X} + [H^T R^{-1} H + S^T P^{-1} S]^{-1} S^T P^{-1} [T - Sx]. \tag{6-13}$$

Equation (6-13) is a convenient and useful form of $\tilde{X}$.

6-4 Linear Estimation with Inequality Constraints

In this section we shall present two methods of obtaining estimates of x. The first method presented is called the Wolfe algorithm for quadratic programming. The second method is more analytical and makes use of the generalized inverse theory. There are many good generalized inverse computer routines for calculating inverses and consequently in many instances it may be desirable to use the second method.

First Method. Again let the observation vector Y be related to the unknown parameter vector x according to the linear model (6-1). Also, let there exist two $n \times 1$ known vectors l and u such that

$$l \leq x \leq u. \tag{6-14}$$

This situation is Case 3 in Section 6-1, where $A = I$. The case where A is a known $m \times n$ matrix of any rank is Exercise 3 of Section 6-5.

If one chooses again the scalar weighted quadratic payoff function

$$Q = (Y - Hx)^T V^{-1}(Y - Hx) \tag{6-15}$$

to minimize with (6-14) as constraints, the problem is immediately recognized as a quadratic programming problem with two-sided inequality constraints.

A formula for $\tilde{X}$ will not be developed in this section; however, a technique for computing the estimate will be formulated in the form of a quadratic programming problem. The estimation technique can be summarized in the following two steps.

Step 1. Rewrite the problem in the form of a quadratic programming problem such that the quantities to be estimated are nonnegative constant parameters. Also, rewrite the inequality constraints into equality constraints.

Step 2. Apply the Kuhn–Tucker theorem so that the linear programming technique that is well known as the simplex method can be used.

Since l and u in (6-14) are known, it is possible to renumber the elements of the parameter vector x with the aim in mind to change x to a nonnegative parameter in the following manner:

$$x = \begin{bmatrix} x_1 \\ x_2 \\ x_3 \end{bmatrix} \quad l = \begin{bmatrix} l_1 \\ l_2 \\ l_3 \end{bmatrix} \quad U = \begin{bmatrix} u_1 \\ u_2 \\ u_3 \end{bmatrix},$$

where x_1 is a vector composed of the nonnegative elements of x, x_2 are the elements of x that are nonpositive, and x_3 are the variables that fit neither of these classifications. Hence

$$\begin{bmatrix} I_1 \\ -I_1 \end{bmatrix} x_1 \leq \begin{bmatrix} u_1 \\ -l_1 \end{bmatrix} \qquad 0 \leq l_1 \leq x_1 \leq u_1 \qquad (6\text{-}16)$$

$$\begin{bmatrix} I_2 \\ -I_2 \end{bmatrix} x_2 \leq \begin{bmatrix} u_2 \\ -l_2 \end{bmatrix} \qquad l_2 \leq x_2 \leq u_2 \leq 0 \qquad (6\text{-}17)$$

$$\begin{bmatrix} I_3 \\ -I_3 \end{bmatrix} x_3 \leq \begin{bmatrix} u_3 \\ -l_3 \end{bmatrix} \qquad l_3 \leq x_3 \leq u_3, \ 0 \leq u_3 \qquad (9\text{-}18)$$

and $0 \leq -l_3$.

Now define $\bar{x}_3$ as the vector composed of the same elements of x_3 but replace all those elements that are negative by zeros; similarly define $-\tilde{x}_3$ as the vector composed of the elements of x_3 but replacing all positive elements by zero. It follows that

$$\begin{bmatrix} \bar{x}_3 \\ \tilde{x}_3 \end{bmatrix} \geq 0$$

and

$$\bar{x}_3 - \tilde{x}_3 = x_3.$$

Finally, we define $\bar{x}_2 = -x_2$ and note that we shall estimate the new vector

$$\bar{x} = \begin{bmatrix} x_1 \\ \bar{x}_2 \\ \bar{x}_3 \\ \tilde{x}_3 \end{bmatrix} \geq \phi. \qquad (6\text{-}19)$$

The constraints (6-16)–(6-18) can now be written as

$$\begin{bmatrix} I_1 \\ -I_1 \end{bmatrix} x_1 \leq \begin{bmatrix} u \\ -l_1 \end{bmatrix} \qquad 0 \leq l_1 \leq u_1 \qquad (6\text{-}20)$$

$$\begin{bmatrix} -I_2 \\ I_2 \end{bmatrix} \bar{x}_2 \leq \begin{bmatrix} u_2 \\ -l_2 \end{bmatrix} \qquad l_2 \leq u_2 \leq 0 \qquad (6\text{-}21)$$

$$\begin{bmatrix} I_3 & \Phi \\ \Phi & I_3 \end{bmatrix} \begin{bmatrix} \bar{x}_3 \\ \tilde{x}_3 \end{bmatrix} \leq \begin{bmatrix} u_3 \\ -l_3 \end{bmatrix}. \qquad (6\text{-}22)$$

The above inequalities can be summarized by

Sec. 6-4 Estimation with Inequality Constraints

and
$$\bar{x} \geq 0$$
where
$$A\bar{x} \leq r,$$

$$A = \begin{bmatrix} I_1 & \Phi & \Phi & \Phi \\ -I_1 & \Phi & \Phi & \Phi \\ \Phi & -I_2 & \Phi & \Phi \\ \Phi & I_2 & \Phi & \Phi \\ \Phi & \Phi & I_3 & \Phi \\ \Phi & \Phi & \Phi & I_3 \end{bmatrix}$$

and

$$r = \begin{bmatrix} u \\ -l_1 \\ u_2 \\ -l_2 \\ u_3 \\ -l_3 \end{bmatrix}.$$

One can now introduce a vector S of unknown "slack variables" such that
$$A\bar{x} + s = r$$
or equivalently
$$[A:I]\begin{bmatrix} \bar{x} \\ s \end{bmatrix} = r, \qquad (6\text{-}23)$$

where I is an identity matrix with the same number of rows as A. If we define

$$x^* = \begin{bmatrix} \bar{x} \\ s \end{bmatrix}$$

$$A^* = [A:I],$$

(6-23) can be rewritten as

$$A^*x^* = r$$

$$x^* \geq \Phi.$$

Now we modify the quadratic payoff function (6-15) by letting

$$H^* = [H_1, -H_2, H_3, -H_3, \Phi],$$

Where H^* and $\bar{x}$ are compatible for multiplication; that is, we can perform the multiplication $H^*\bar{x}$. Then the linear model (6-1) is modified to be

$$Y = H^*x^* + Z \qquad (6\text{-}24)$$

and (6-15) is modified to be

$$Q^* = (y - H^*x^*)^T V^{-1}(y - H^*x^*)$$

or

$$Q^* = y^T V^{-1}y - 2x^{*T}H^{*T}V^{-1}y + x^{*T}H^{*T}V^{-1}H^*x^* \qquad (6\text{-}25)$$

subject to the constraints

$$A^*x^* = r$$
$$x^* \geq \Phi. \qquad (6\text{-}26)$$

Equation (6-25) along with constraints (6-26) can now be solved by quadratic programming to yield a solution $\hat{x}$, which will minimize the "payoff" function defined by (6-7).

Our purpose in this text is not to discuss all the different techniques for solving quadratic programming problems; however, we shall introduce a technique for transforming the problem described above to a linear program. Theorem 6-4.1 describes how this can be done.

THEOREM 6-4.1. (*Kuhn–Tucker Theorem*). *The vector* $\hat{x}$ *solves the quadratic program problem; that is,* $\hat{x}$ *maximizes*

$$-Q = -y^T V^{-1}y + 2x^{*T}H^*V^{-1}y - \tfrac{1}{2}x^{*T}(2H^{*T}V^{-1}H^*)x^* \qquad (6\text{-}27)$$

subject to

$$A^*x^* = r \qquad (6\text{-}28)$$

$$x^* \geq \Phi \qquad (6\text{-}29)$$

if and only if x *satisfies (6-28) and (6-29) and there exist vectors* u $\geq \Phi$ *and* w *such that*

$$2H^{*T}V^{-1}H^*x^* - u + A^{*T}w - 2H^{*T}V^{-1}y = 0 \qquad (6\text{-}30)$$

and

$$u^T x^* = 0. \qquad (6\text{-}31)$$

Sec. 6-4 Estimation with Inequality Constraints

The proof of Theorem 6-4.1 can be found in Reference [27]. In solving the problem described in the theorem, one must find $\hat{x}$, which satisfies (6-28)–(6-31). One such method for finding $\hat{x}$ is called the Wolfe algorithm for quadratic programming. This method is similar to simplex programming and easily adaptable to a computer. To illustrate, let us consider the following equations:

$$2H^{*T}V^{-1}H^*x^* - u + A^*w = 2H^{*T}V^{-1}y$$

$$A^*x^* = r.$$

The condition $u^T x^* = \Phi$ will be incorporated into the program later. In the normal simplex procedure, slack variables are added to each equation in the system; that is,

$$2H^{*T}V^{-1}H^*x^* - u + A^*w + m = 2H^*TV^{-1}y$$

$$A^*x^* + t = r.$$

We can now use the simplex algorithm [198] to minimize m and t to Φ. Since we are assuming there does exist a vector $\hat{x}$ that fills the first conditions of Theorem 6-4.1, then the minimum of both m and t must equal Φ, thus giving us an $\hat{x}$, which satisfies

$$2H^*V^{-1}H^*x^* - u + A^*w = 2H^{*T}V^{-1}y$$

$$A^*r^* = r.$$

To fill the condition $u^T x^* = 0$, a slight alteration is made in the simplex algorithm. This alteration can be summarized as follows:

1. Do not allow u_i into the nonzero solution set at any stage, unless x_i is not contained in the nonzero solution set.
2. In like fashion, do not admit x_i to the nonzero solution set unless $u_i = \Phi$.

Under these conditions, $u^T x^* = 0$, since either $u_i = \Phi$ or $x_i = \Phi$, for each i. To illustrate consider the following Example.

Example: Suppose we are given the quadratic form

$$Q = \tfrac{5}{2} + (x_1, x_2, x_3)\begin{pmatrix}1\\0\\-2\end{pmatrix} + \tfrac{1}{2}(x_1, x_2, x_3)\begin{pmatrix}1 & 0 & 0\\0 & 1 & 0\\0 & 0 & 1\end{pmatrix}\begin{pmatrix}x_1\\x_2\\x_3\end{pmatrix}$$

and we wish to minimize Q subject to the constraints

$$(1, -1, 1)\begin{pmatrix} x_1 \\ x_2 \\ x_3 \end{pmatrix} = 1, \qquad \begin{pmatrix} x_1 \\ x_2 \\ x_3 \end{pmatrix} \geq \Phi.$$

We first note that

$$2H^{*T}H^* = \begin{pmatrix} 1 & 0 & 0 \\ 0 & 1 & 0 \\ 0 & 0 & 1 \end{pmatrix}, \qquad A^* = (1, -1, 1),$$

and

$$2H^*y = \begin{pmatrix} -1 \\ 0 \\ 2 \end{pmatrix},$$

so that

$$\begin{pmatrix} 1 & 0 & 0 \\ 0 & 1 & 1 \\ 0 & 0 & 1 \end{pmatrix}\begin{pmatrix} x_1 \\ x_2 \\ x_3 \end{pmatrix} - \begin{pmatrix} 1 & 0 & 0 \\ 0 & 1 & 0 \\ 0 & 0 & 1 \end{pmatrix}\begin{pmatrix} u_1 \\ u_2 \\ u_3 \end{pmatrix} + \begin{pmatrix} 1 \\ -1 \\ 1 \end{pmatrix}w - \begin{pmatrix} -1 \\ 0 \\ 2 \end{pmatrix} = \Phi$$

$$(u_1 \; u_2 \; u_3)\begin{pmatrix} x_1 \\ x_2 \\ x_3 \end{pmatrix} = 0.$$

Therefore, the initial tableau is

x_1	x_2	x_3	u_1	u_2	u_3	w	m_{11}	m_{12}	m_{13}	m_{21}	m_{22}	m_{23}	t	
1*	−1	1	0	0	0	0	0	0	0	0	0	0	1	1
1	0	0	−1	0	0	1	1	0	0	−1	0	0	0	−1
0	1	0	0	−1	0	−1	0	1	0	0	−1	0	0	0
0	0	1	0	0	−1	1	0	0	1	0	0	−1	0	2
1	1	1	1	1	1	1	0	0	0	0	0	0	0	

The solutions at each stage are as follows:

Sec. 6-4   Estimation with Inequality Constraints

Basic Solutions	x_1	x_2	x_3	u_1	u_2	u_3	w	m_{11}	m_{12}	m_{13}	m_{21}	m_{22}	m_{23}	T
0									0	2			1		1
1	1								0	2			2		0
2		1						2		2	4				
3				1				1		1			2		
4		$\frac{1}{2}$	$\frac{3}{2}$					$\frac{1}{2}$					$\frac{3}{2}$		
5		$\frac{1}{2}$	$\frac{3}{2}$	$\frac{3}{2}$				$\frac{1}{2}$							

Thus the final solution is

$$\begin{pmatrix} x_1 \\ x_2 \\ x_3 \end{pmatrix} = \begin{pmatrix} 0 \\ \frac{1}{2} \\ \frac{3}{2} \end{pmatrix}.$$

The solutions above were achieved by the following method:

1. A bottom row is added to the initial tableau consisting of 1's below the x's, u's, and w, and of 0's below the m's and t. We are willing to enter into the solution set the variables with 1's underneath.
2. The pivot element is chosen by taking a column underneath the variables that may enter into the solution set and dividing its elements into the corresponding elements of the extreme right column. The row with the least positive (nonzero) quotient is selected.
3. Linear row transformations are made on the tableau until a column with 1 at the pivot and 0's elsewhere in the column is obtained. We choose to work with the first column and the second tableau is then

x_1	x_2	x_3	u_1	u_2	u_3	w	m_{11}	m_{12}	m_{13}	m_{21}	m_{22}	m_{23}	T
1	−1	1	0	0	0	0	0	0	0	0	0	1	1
0	1	−1	1	0	0	1	1	0	0	−1	0	−1	−2
0	1	0	0	−1	0	−1	0	1	0	0	−1	0	0
0	0	1	0	0	−1	1*	0	0	1	0	0	−1	2
0	2	0	1	1	1	1							

110 On Linear Estimation with Constraints Chap. 6

4. Since x_1 is in the solution set, we eliminate u_1 as a possible entry. x_2 is our next logical choice, but since all the quotients we obtained by dividing the elements of the column underneath x_2 into the corresponding elements on the extreme right column are zero or negative, then we disregard this column. A similar argument can be given to eliminate columns underneath x_3, u_1, u_2, and u_3. We choose to work with the column underneath w. The pivot element will be at the starred position.
5. The process is continued in this fashion, never entering an m or t variable, until there are no m or t variables in the solution. Once this has been accomplished, a minimal point occurs. Thus we can read our solution at this point.
6. The solution we read from our first tableau is $x_1 = 0$, $m_{12} = 0$, $m_{13} = 2$, $m_{21} = 2$ with all unspecified variables equal to zero. The solution variables always have unit matrix columns beneath them. Continuing this process, we arrive at a minimum solution. The third, fourth, and fifth tableau are as follows:

x_1	x_2	x_3	u_1	u_2	u_3	w	m_{11}	m_{12}	m_{13}	m_{21}	m_{22}	m_{23}	T
1	−1	1*	0	0	0	0	0	0	0	0	0	1	1
0	1	−2	1	0	2	0	1	0	−1	−1	0	1	1	−2
0	1	1	0	−1	−1	0	0	1	1	0	−1	−1	−1	1
0	0	1	0	0	−1	1	0	0	1	0	0	−1	−1	1
0	2	−1	1	1	2								

x_1	x_2	x_3	u_1	u_2	u_3	w	m_{11}	m_{12}	m_{13}	m_{21}	m_{22}	m_{23}	T
1	−1	1	0	0	0	0	0	0	0	0	0	1	1
2	−1	0	1	0	2	0	1	0	−1	−1	0	1	1	−2
0	2*	0	0	−1	−1	0	0	1	1	0	−1	−1	−1	1
0	1	0	0	0	−1	1	0	0	1	0	0	−1	−1	1
1	1	0	1	1	2	0							

Sec. 6-4   Estimation with Inequality Constraints   111

x_1	x_2	x_3	u_1	u_2	u_3	w	m_{11}	m_{12}	m_{13}	m_{21}	m_{22}	m_{23}	T
1	0	1	0	$-\frac{1}{2}$	$-\frac{1}{2}$	0	0	$\frac{1}{2}$	$\frac{1}{2}$	0	$-\frac{1}{2}$	$-\frac{1}{2}$	0	$\frac{3}{2}$
2	0	0	1*	$-\frac{3}{2}$	$\frac{3}{2}$	0	1	$\frac{1}{2}$	$-\frac{1}{2}$	-1	$-\frac{1}{2}$	$\frac{1}{2}$	$\frac{1}{2}$	$-\frac{3}{2}$
0	1	0	0	$-\frac{1}{2}$	$-\frac{1}{2}$	0	0	$\frac{1}{2}$	$\frac{1}{2}$	0	$-\frac{1}{2}$	$-\frac{1}{2}$	$-\frac{1}{2}$	$\frac{1}{2}$
0	0	0	0	$\frac{1}{2}$	$-\frac{1}{2}$	1	0	$-\frac{1}{2}$	$\frac{1}{2}$	0	$\frac{1}{2}$	$-\frac{1}{2}$	$-\frac{1}{2}$	$\frac{1}{2}$
1	0	0	1	$\frac{1}{2}$	$\frac{5}{2}$	0								

The above solution gives a minimum to

$$Q^T Q = \tfrac{5}{2} + (x_1, x_2, x_3)\begin{pmatrix}1\\0\\-2\end{pmatrix} + \tfrac{1}{2}(x_1, x_2, x_3)\begin{pmatrix}1 & 0 & 0\\0 & 1 & 0\\0 & 0 & 1\end{pmatrix}\begin{pmatrix}x_1\\x_2\\x_3\end{pmatrix}.$$

This minimum occurs at

$$\begin{pmatrix}\hat{x}_1\\\hat{x}_2\\\hat{x}_3\end{pmatrix} = \begin{pmatrix}0\\\tfrac{1}{2}\\\tfrac{3}{2}\end{pmatrix}.$$

Second Method. For our discussion we shall assume the constraints to be of the form

$$Ax \leq t, \qquad (6\text{-}32)$$

where A is a known $p \times n$ matrix of any rank and t is a $p \times 1$ known vector. The first step is to add a $p \times 1$ vector u of nonnegative slack variables to the left-hand side of (6-32), yielding the equivalent condition

$$Ax + u = t, \qquad u \geq 0.$$

By defining

$$x^* = \begin{bmatrix}x\\u\end{bmatrix}, \qquad A^* = [A, I]$$

and

$$H^* = [H, 0],$$

we can formulate the general restricted linear model as

$$Y = H^* x^* + Z \tag{6-33}$$

with the linear constraints

$$A^* x^* = t, \quad u \geq 0,$$

where

$$x^* = [x^T, u^T]^T.$$

Note that the matrix A^* is of full-row rank, but that the matrix H^* is not of full-column rank. It can be shown (see Exercise 6) that the minimum variance linear unbiased estimator for x in (6-1) is

$$\tilde{X} = A^+ t + C^+ H^T V^{-1} (Y - H A^+ t) \tag{6-34}$$

with

$$C = (I - A^+ A) H^T V^{-1} H (I - A^+ A),$$

where H is any rank and A is full-row rank. The corresponding covariance of $\tilde{X}$ is C^+. Note that when H is full column, (6-34) reduces to (6-6).

With (6-34) and ignoring the p-nonnegativity conditions for the moment, we can find an estimator $\tilde{X}^*$ that is "best" among all estimators of x^* that are linear combinations of Y and t.

Clearly, if $\tilde{X}^*$ satisfies the nonnegativity conditions (i.e., $\tilde{X}_i^* \geq 0$, $i = n+1, \ldots, n+p$), then this is the desired estimator. In general, however, this will not be the case. The following discussion provides a method for treating the case of negativitiy among some of the last p entries of $\tilde{X}^*$.

We first reformulate the linear regression problem as follows:
Find the vector $\hat{x}^*$ that minimizes

$$\psi(x^*) = (y - H^* x^*)^T V^{-1} (y - H^* x^*) \tag{6-35}$$

subject to the constraints

$$A^* x^* = t, \quad x_i^* \geq 0, \quad i = n+1, \ldots, n+p. \tag{6-36}$$

Then $\hat{x}^*$ will be the best linear estimator of x^* from the class of those that satisfy (6-36). If we ignore the nonnegativity constraints, then the vector estimate $\tilde{x}^*$ that minimizes (6-35) subject to $A^* \tilde{x}^* = t$ is given by

$$x^* = A^{*+} t + C^+ H^{*T} V^{-1} (Y - H^* A^{*+} t), \tag{6-37}$$

where

$$C = (I - A^{*+} A^*) H^{*T} V^{-1} H^* (I - A^{*+} A^*).$$

Sec. 6-4 Estimation with Inequality Constraints

Define

$$U(x) = \{i \mid n+1 \leq i \leq n+p, \quad x_i = 0\}$$

$$V(x) = \{i \mid n+1 \leq i \leq n+p, \quad x_i \leq 0\}.$$

We now prove the following.

THEOREM 6-4.2. *For any two vectors* x *and* $\underline{x}$

$$\psi(z) \leq \max\{\psi(x), \psi(\underline{x})\}$$

for every $z = \theta x + (q - \theta)\underline{x}, 0 < \theta < 1$. *If* $\psi(x) \neq \psi(\underline{x})$, *then strict inequality holds.*

Proof: $\psi(x^*)$ given in (6-35) is clearly convex; that is, for every θ, $p < \theta < 1$ and any two vectors x and $\underline{x}$,

$$\psi[\theta x + (1-\theta)\underline{x}] \leq \theta \psi(x) + (1+\theta)\psi(\underline{x}).$$

Hence,

$$\psi[\theta x + (1-\theta)\underline{x}] \leq \psi(x) + (1-\theta)[\psi(\underline{x}) - \psi(x)]$$

$$= \psi(\underline{x}) + \theta[\psi(x) - \psi(\underline{x})].$$

The results follow immediately.

THEOREM 6-4.3. *If* $V(\tilde{x}^*) \neq \phi$ *for the vector estimate* $\tilde{x}^*$ *of Equation (6-37), then there exists a vector estimate* $\hat{x}^*$, *which minimizes (6-35) subject to the constraints (6-36) such that*

$$V(\tilde{x}^*) \cap U(\hat{x}^*) \neq \phi.$$

The vector $\hat{X}^*$ *is then the best linear (in y and t) estimator of* x^* *in model (6-35), with restriction (6-36).*

Proof: Obviously, $\psi(\tilde{x}^*) \leq \psi(\hat{x}^*)$. If $\psi(\tilde{x}^*) < \psi(\hat{x}^*)$, then suppose that $U(\hat{x}^*) \cap V(\tilde{x}^*) = \phi$. Then $\hat{x}_h^* > 0$ and $\tilde{x}_h^* < 0$ for each $h \in V(\tilde{x}^*)$. Thus, for each such h, there exists a real number θ_h, $0 < \theta_h < 1$, such that

$$\theta_h \hat{x}_h^* + (1 - \theta_h)\tilde{x}_h^* = 0.$$

Define $\tilde{\theta} = \max_{h \in V(\tilde{x}^*)} \{\theta_h\}$. Then $0 < \tilde{\theta} < 1$, and we have

$$\tilde{z} = \tilde{\theta}\hat{x}^* + (1 - \tilde{\theta})x^* \geq 0,$$

$$A^*\tilde{z} = \tilde{\theta}A^*\hat{x}^* + (1 - \tilde{\theta})A\tilde{x}^* = t.$$

By Theorem 6-4.2.,

$$\psi(\tilde{z}) < \max \{\psi(\tilde{x}^*), \psi(\hat{x}^*)\} = \psi(\hat{x}^*).$$

But this contradicts the properties of $\hat{x}^*$. Hence we must conclude that $U(\hat{x}^*) \cap V(\tilde{x}^*) \neq \phi$.

Suppose next that $\psi(\tilde{x}^*) = \psi(\hat{x}^*)$. Then the vector z constructed above satisfies $\tilde{z} \geq 0$, $A^*\tilde{z} = t$, $\psi(\tilde{z}) = \psi(\hat{x}^*)$, and $U(\tilde{z}) \cap V(\tilde{x}^*) \neq \phi$, since

$$\tilde{z}_h = \tilde{\theta}\hat{x}_h + (1 - \tilde{\theta})\tilde{x}_h^* = 0$$

for some $h \in V(\tilde{x}^*)$. Thus $\tilde{z}$ serves as the required solution.

It is well known that minimization of (6-35) yields the estimator with the desired properties [1, 184]. The fact that some of the entries $\hat{x}_h^*$, $h = n + 1, \ldots, n + p$, must be zero merely reflects the fact that in the desired estimator, the corresponding linear inequality constraint is satisfied as an equality. This completes the Proof.

It is now possible to describe an algorithm for finding the desired minimum variance linear estimator.

Step 1. Calculate $\tilde{x}^*$ as given in (6-37) and list the set $V(\tilde{x}^*)$, if nonempty.

Step 2. One by one, set the entries of x^* corresponding to the elements of $V(\tilde{x}^*)$ equal to zero and repeat Step 1, each time solving an $(n + p - 1)$-dimensional system. This can be done by crossing out the hth element of x^*, the hth column of A^*, and the hth column of H^* as h ranges over $V(\tilde{x}^*)$.

Step 3. For each resulting vector in Step 2 that does not satisfy (6-36), set its other components equal to zero, one at a time, and solve the remaining $(n + p - 2)$-dimensional system. However, at no time should a set of components be set equal to zero when some proper subset (when set equal to zero) yields an estimate that satisfies (6-36).

Step 4. Repeat Step 3 for all triples of which at least one member yielded no solution in Step 2, and no subset of which yielded a solution in Step 3. Do the same for 4-tuples, etc.

After a finite number of steps, the sought-for estimator can be obtained by noting which estimator derived in the algorithm minimizes $\psi(x)$. If any two estimators $\underline{x}$ and $\underline{\underline{x}}$ yield an identical minimum, then any vector of the form

$$z = \theta\underline{x} + (1 - \theta)\underline{\underline{x}},$$

$0 \leq \theta \leq 1$, will also have the desired properties, by Theorem 6-4.2.

Sec. 6-4 Estimation with Inequality Constraints

In the special case where the matrix H of the original model 1 (see p.) has all its columns linearly independent, certain arguments given in Reference [181] apply. The most important is that if an estimate $\hat{x}$ violates none of the given constraints, then $\hat{x}$ is the desired estimate if and only if removing the zero restriction on any zero element of $\hat{x}$ results in an estimate whose corresponding element is negative. Theorem 6-4.2 provides a very powerful addition to the proposed algorithm for the full-rank case, but it does not hold in general.

Numerical Example: The following example, originally due to Telser [181], has also been treated in References [94], [113], and [185]. The problem is concerned with finding the best estimates for the entires of a transition probability matrix in a finite Markov process. Telser's example lists the sales from 1925 to 1943 of the three leading brands of cigarettes for that period. It will be assumed that the linear model takes the form

$$y^{(j)} = H^{(j)} p^{(j)} + Z^{(j)}, \quad j = 1, 2, 3,$$

where the transition probability matrix P is given by

$$p = [p^{(1)}, p^{(2)}, p^{(3)}],$$

and y_{jt} is the proportion of smokers of brand j at time t, $H^{(j)}$ consists of data from time $t - 1$, and $Z^{(j)}$ is a random vector such that $E[Z^{(j)}] = 0$ and $E[Z^{(j)} Z^{(j)T}] = I$. The estimates $\hat{P}$ for the transposition probability matrix must satisfy the constraints

$$\sum_{j=1}^{3} p_{ij} = 1, \quad i = 1, 2, 3,$$

and

$$p_{ij} \geq 0, \quad i, j = 1, 2, 3.$$

Note that the constraint $1 \geq p_{ij}$ is implicit in the above and need not be imposed.

In the cigarette example,

$$H^{(1)} = H^{(2)} = H^{(3)} = \begin{bmatrix} 2.65847323 & 2.19322173 & 1.98780504 \\ 2.19322173 & 2.06889281 & 1.69718546 \\ 1.98780504 & 1.69718546 & 1.51620950 \end{bmatrix}$$

and

$$[y^{(1)}, y^{(2)}, y^{(3)}] = \begin{bmatrix} 2.57646345 & 2.28265139 & 1.98038516 \\ 2.13442225 & 2.11107622 & 1.70480153 \\ 1.94161430 & 1.75027239 & 1.50931331 \end{bmatrix}.$$

The linear model can be expressed as

$$\begin{bmatrix} Y^{(1)} \\ Y^{(2)} \\ Y^{(3)} \end{bmatrix} = \begin{bmatrix} H^{(1)} & 0 & 0 \\ 0 & H^{(2)} & 0 \\ 0 & 0 & H^{(3)} \end{bmatrix} \begin{bmatrix} p^{(1)} \\ p^{(2)} \\ p^{(3)} \end{bmatrix} + \begin{bmatrix} Z^{(1)} \\ Z^{(2)} \\ Z^{(3)} \end{bmatrix},$$

$$E[Z^{(i)}] = 0, \quad E[Z^{(i)}Z^{(i)T}] = I, \quad i = 1, 2, 3,$$

subject to the constraints

$$Ap = t, \quad p \geq 0,$$

with $A = [I, I, I]$, $p = [p^{(1)T}, p^{(2)T}, p^{(3)T}]^T$, and $t = [1, 1, 1]^T$.

Ignoring the nonnegativity constraints for the moment, we can calculate $\tilde{p}$ from

$$\tilde{p} = A^+ t + G^+(y - HA^+ t),$$

where $G = H(I - A^+ A)$. This yields

$$\tilde{P} = \begin{bmatrix} .5453 & .3575 & .0972 \\ -.0744 & .9654 & .1090 \\ .6489 & -.3949 & .7460 \end{bmatrix} \quad (6\text{-}38)$$

This estimate satisfies all of the linear constraints except two of the nonnegativity constraints. According to the algorithm developed, the admissible estimate $\hat{p}$ of p must satisfy at least one of the constraints

$$\hat{p}_{21} = 0, \quad \hat{p}_{32} = 0.$$

Setting $\hat{p}_{21} = 0$ and solving the remaining 8-dimensional problem yields

$$\hat{p}^{(2,1)} = \begin{bmatrix} .6847 & .2878 & .0275 \\ 0 & .9282 & .0718 \\ .3809 & -.2609 & .8800 \end{bmatrix}. \quad (6\text{-}39)$$

This estimate is clearly inadmissible because of the negative element in the (3, 2) position.

Continuing with the algorithm, we set $\hat{p}_{32} = 0$ instead of $\hat{p}_{21}$ and obtain

$$\hat{P}^{(3,2)} = \begin{vmatrix} .6580 & .1321 & .2099 \\ -.0318 & .8802 & .1516 \\ .4514 & 0 & .5486 \end{vmatrix} \quad (6\text{-}40)$$

The next step is to set both $\hat{p}_{21}$ and $\hat{p}_{32}$ equal to zero simultaneously.

This yields the result

$$\hat{P}^{(2,1),(3,2)} = \begin{bmatrix} .7089 & .1364 & .1547 \\ 0 & .8751 & .1249 \\ .3484 & 0 & .6516 \end{bmatrix} \quad (6\text{-}41)$$

This estimate is admissible. Indeed, since the matrix H in the model is nonsingular, (6-41) is the sought-for estimate, by virtue of the fact that in $\hat{P}^{(2,1)}$, $p_{32} < 0$, and in $\hat{P}^{(3,2)}$, $p_{21} < 0$.

EXERCISES

1. Let H and A be $m \times n$ and $p \times n$ matrices, respectively, such that $r(H) = n$ and $r(A) = p$. Let V^{-1} be a symmetric positive definite matrix. Then show that $[A(H^T V^{-1} H)^{-1} A^T]^{-1}$ exists.

2. Let the linear model be (6-1). Let the linear constraints on x be

$$Ax = t,$$

   where A and t are known $p \times n$ and $p \times 1$ matrices, respectively, such that $r(A) = q < \min(p, n)$. Can a minimum variance estimator be determined? If the answer is yes, find the estimator.

3. Let the linear model be (6-1). Estimate x, where it is know that the following inequalities are true:

$$l \leq Ax \leq u,$$

   where l and u are known nonrandom $m \times 1$ vectors and A is an $m \times n$ matrix.

4. With the aid of Exercises 5, 6, and 7, prove Theorem 6-4.1.

5. Let

$$Q = 5 + (x_1, x_2, x_3) \begin{pmatrix} 1 \\ 3 \\ 2 \end{pmatrix} + (x_1, x_2, x_3) \begin{pmatrix} 2 & 0 & 0 \\ 0 & 4 & 0 \\ 0 & 0 & 6 \end{pmatrix} \begin{pmatrix} x_1 \\ x_2 \\ x_3 \end{pmatrix},$$

   where

$$0 \leq (1, 2, 1) \begin{pmatrix} x_1 \\ x_2 \\ x_3 \end{pmatrix} \leq 10, \quad \begin{pmatrix} x_1 \\ x_2 \\ x_3 \end{pmatrix} \geq \phi.$$

   Find x_1, x_2, and x_3 such that Q is minimized.

6. In the linear model (6-1) let H be of any rank. Let $Ax = t$, where A is full-row rank. Show that the minimum variance linear unbiased estimator of x is

$$\hat{X} = A^+ t + C^+ H^T V^{-1}(Y - HA^+ t)$$

with

$$C = (I - A^+ A) H^T V^{-1} H (I - A^+ A).$$

Show that Cov $(\hat{X}) = C^+$.

7. If $E[ZZ^T] = \sigma^2 I$ in the linear model (6-1) and $Ax = t$ with A full-row rank, show that $\hat{X} = A^+ t + G^+(Y - HA^+ t)$, where $G = H(I - A^+ A)$.

CHAPTER 7

Estimating a Stochastic Process in a Dynamic Model Using Continuous Data

7-1 Introductory Remarks

In this chapter we shall extend our theory of linear estimation to a class of problems in which the $n \times 1$ parameter vector is composed of elements that are continuous real-valued functions of a real independent variable t (time). For clarity, we list several definitions that not only will be basic to our discussion but will also relate what is discussed here to the existing technical literature.

Definition 7-1.1. *Let* T *be any set and let* X(t) *be a random variable for each* t *belonging to* T. *Then we define a random or stochastic process as the collection of random variables* $\{X(t); t \in T\}$.

Definition 7-1.2. *Let* $\{X(t); t \in T\}$ *be a stochastic process. Then for each fixed* t *the* E[X(t)] *is given by*

$$E[X(t)] = \int_{-\infty}^{\infty} x(t) f_{X(t)}[x(t)] \, dx(t), \qquad (7\text{-}1)$$

where $f_{X(t)}[x(t)]$ *denotes the probability density function of* X(t).

Definition 7-1.3. *Let $\{X(t); t \in T\}$ be a stochastic process with $X(t)$ a continuous random variable for each $t \in T$. Then for any t_0 and $t_1 \in T$ we define*

1. *the autocorrelation $R_x(t_0, t_1)$ by*

$$R_x(t_0, t_1) = E[X(t_0)X(t_1)] \tag{7-2}$$

and

2. *the autocovariance or covariance kernel, $C_x(t_0, t_1)$ by*

$$C_x(t_0, t_1) = E\{[X(t_0) - u(t_0)][X(t_1) - u(t_1)]^T\}, \tag{7-3}$$

where $u(t_0)$ and $u(t_1)$ are $E[X(t_0)]$ and $E[X(t_1)]$, respectively.

From the definition of expectation it follows from (7-2) and (7-3) that

$$C_x(t_0, t_1) = R_x(t_0, t_1) - u(t_0)u(t_1) \tag{7-4}$$

Definition 7-1.4. *Let $\{X(t); t \in T\}$ and $\{Y(t); t \in T\}$ be stochastic processes and t_0 and t_1 be any two points in T. Then the crosscorrelation $R_{XY}(t_0, t_1)$ and crosscovariance $C_{XY}(t_0, t_1)$ are defined by*

$$R_{XY}(t_0, t_1) = E[X(t_0)Y(t_1)]$$

$$C_{XY}(t_0, t_1) = E\{[X(t_0) - u_X(t_0)][Y(t_1) - u_Y(t_1)]\}$$

$$= R_{XY}(t_0, t_1) - u_X(t_0)u_Y(t_1).$$

7-2 The Dynamic Linear Model

In Chapter 5 we considered estimating the nonrandom $n \times 1$ vector-valued parameter function $x(t)$ in the linear model for continuous data

$$Y(t) = h(t)x(t) + V(t), \quad t_0 < t < t_1, \tag{7-5}$$

where $Y(t)$ is a $p \times 1$ vector stochastic process of observations and $h(t)$ is a known matrix-valued function relating $Y(t)$ and $x(t)$. The error stochastic process $V(t)$ is unobservable, but it is known that

$$E[V(t)] = \phi \quad \text{for all } t \tag{7-5a}$$

and

$$E[V(t)V^T(s)] = R(t)\delta(t - s), \quad \text{for every } t \text{ and } s, \tag{7-5b}$$

Sec. 7-2 The Dynamic Linear Model

where $\delta(t - s)$ denotes the Dirac delta functional. We list for easy reference a property of $\delta(t - s)$; that is, if $g(s)$ is a continuous function, then

$$\int_{-\infty}^{\infty} \delta(t - s) g(t)\, dt = g(s).$$

In this chapter we require $X(t)$ to be a random vector such that

$$Y(t) = h(t) X(t) + V(t). \tag{7-6}$$

We also require that conditions (7-5a) and (7-5b) be true along with the following known additional conditions:

$$E[X(t)] = \phi$$

$$E[X(t_0)X(t_0)^T] = P(t_0)$$

and

$$E[X(t)V^T(s)] = \phi, \quad \text{for every } t \text{ and } s.$$

Definition 7-2.1. *The linear model defined by* (7-6) *and the differential equation constraint*

$$\frac{dX(t)}{dt} = F(t)X(t) + G(t)U(t) \tag{7-7}$$

define a dynamic linear model. The matrix-valued functions F(t) *and* G(t) *are known, and* U(t) *is a random process that cannot be observed but it is known that*

$$E[U(t)] = \phi$$

$$E[U(t)U^T(s)] = Q(t)\delta(t - s)$$

$$E[U(t)X^T(s)] = \phi$$

$$E[U(t)V^T(s)] = \phi, \quad \text{for every t and s.}$$

The problem remains the same. We wish to estimate $X(t)$ using the observation vector $Y(t)$, $t_0 \leq t \leq t_1$, by a linear estimator

$$\hat{X}(t) = \int_{t_0}^{t_1} A(t, s) Y(s)\, ds. \tag{7-8}$$

We shall select $A(t, s)$ so that the expected mean-square loss function

$$E\{[\hat{X}(t) - X(t)][\hat{X}(t) - X(t)]^T\}$$

is minimized. If $\hat{X}(t)$ is unbiased, it is then minimum variance.

For our purposes it is assumed that all functions involved will satisfy enough regularity conditions so that we may interchange the order of integration, and the order of integration and expectation, and so that we can differentiate under the integral sign.

To make our discussion self-contained it seems desirable to list and prove several theorems cited again and again in the literature.

LEMMA 7-2.1. (*The Orthogonal Projection Theorem*). *Let* U *be a subspace of the abstract linear inner product space* K, *then given a vector* $x \in K$, *there is a vector* $u_0 \in U$ *such that for all* $u \in U$

$$\|x - u_0\| \leq \|x - u\|$$

if and only if the inner product

$$\langle x - u_0, u \rangle = 0 \qquad (7\text{-}9)$$

for every $u \in U$. *Moreover, if there exists another* u_1 *satisfying* (7-9), *then*

$$\|u_0 - u_1\| = 0.$$

Proof: Consider

$$\|x - u\|^2 = \|x - u_0 + u_0 - u\|^2$$
$$= \|x - u_0\|^2 + \|u - u_0\|^2 + 2\langle x - u_0, u_0 - u \rangle.$$

Now suppose that (7-9) holds; then

$$\|x - u\|^2 = \|x - u_0\|^2 + \|u - u_0\|^2$$

and finally,

$$\|x - u\|^2 \geq \|x - u_0\|^2.$$

which implies that

$$\|x - u\| \geq \|x - u_0\|$$

for every $u \in U$.

Conversely, suppose that

$$\|x - u\| \geq \|x - u_0\|$$

for every $u \in U$. Now if there exists a vector $u_1 \in U$ such that

$$\langle x - u_0, \quad u_1 \rangle = \alpha$$

and

$$\alpha \neq 0,$$

then

$$\|x - u_0 - \beta u_1\|^2 = \|x - u_0\|^2 + \beta^2 \|u_1\|^2 + 2\langle x - u_0, -\beta u_1\rangle$$
$$= \|x - u_0\|^2 + \beta^2 \|u_1\|^2 - 2\alpha\beta.$$

Now the value of β can be chosen so that

$$\|x - u_0 - \beta u_1\|^2 \leq \|x - u_0\|^2,$$

which contradicts that u_0 is the vector that minimizes $\|x - u\|$. The theorem is proved.

THEOREM 7-2.1. *(The Wiener–Hopf Equation). A necessary and sufficient condition for* $\hat{X}(t)$ *in (7-8) to be a minimum variance estimator for* X(t) *is that the matrix* A(t, s) *satisfy the relation*

$$\text{Cov}\,[X(t), Y(w)] - \int_{t_0}^{t_1} A(t, s)\,\text{Cov}\,[Y(s), Y(w)]\,ds = 0, \qquad (7\text{-}10)$$

or equivalently,

$$\text{Cov}\,[\tilde{X}(t), Y(w)] = 0 \qquad (7\text{-}11)$$

for every $t_0 \leq w \leq t_1$, *where*

$$\tilde{X}(t) = X(t) - \hat{X}(t).$$

Proof: Let Z denote the set of all $n \times 1$ random vector functions $X(t)$ with zero means and finite variances. Let T denote the set of all $n \times 1$ random vectors defined by

$$X^1(t) = \int_{t_0}^{t_1} B(t, s)\,Y(s)\,ds,$$

where $B(t, s)$ is a $p \times n$ matrix of continuously differentiable functions in both arguments, t and s. We define an inner product (see Definition 1-1.2) as

$$\langle X^1(t), X(t)\rangle = E[X^1(t) X^T(t)]. \qquad (7\text{-}12)$$

Suppose that Cov $[\tilde{X}(t), Y(w)] = \phi$ for every w in the interval $[t_0, t_1]$. Then

$$\langle \tilde{X}(t), X^1(t)\rangle = E[\tilde{X}(t) X^1(t)^T]$$
$$= E\left[\tilde{X}(t) \int_{t_0}^{t_1} Y^T(s) B^T(t, s)\,ds\right]$$
$$= \int_{t_0}^{t_1} \text{Cov}\,[\tilde{X}(t), Y(s)] B^T(t, s)\,ds.$$

But, by (7-11), $\text{Cov}[X(t), Y(s)] = 0$ for every s in the interval $[t_0, t_1)$, which implies

$$\langle \tilde{X}(t), X^1(t) \rangle = 0.$$

Recall that $\tilde{X}(t) = X(t) - \hat{X}(t)$ and by Lemma 7-2.1, $\hat{X}(t)$ is a minimum variance estimator of $X(t)$ with respect to the set T.

Conversely, suppose $A(t, s)$ has been selected such that $\hat{X}(t)$ is a minimum variance estimator of $X(t)$ with respect to the set T. Then

$$\langle \tilde{X}(t), X^1(t) \rangle = 0$$

for every $X^1(t) \in T$. We can write

$$\langle \tilde{X}(t), X^1(t) \rangle = \int_{t_0}^{t_1} \text{Cov}[\tilde{X}(t), Y(s)] B^T(t, s) \, ds = 0$$

for every $X^1(t) \in T$. Now define

$$B(t, s) = \text{Cov}[\tilde{X}(t), Y(s)].$$

Hence

$$\langle \tilde{X}(t), X^1(t) \rangle = \int_{t_0}^{t_1} B(t, s) B^T(t, s) \, ds = \phi.$$

But this is possible only if $B^T(t, s) \equiv \phi$, since for every s, $B(t, s) B^T(t, s)$ is nonnegative definite and, on integrating the diagonal elements, it will not vanish. Hence it follows that $\langle \tilde{X}(t), X^1(t) \rangle = 0$ implies that $\text{Cov}[\tilde{X}(t), Y(s)] = 0$ for every s in the interval $[t_0, t_1)$.

COROLLARY 7-2.1. $E[\tilde{X}(t) \hat{X}^T(t)] = 0$.

Proof: Since $\hat{X}(t) = \int_{t_0}^{t_1} A(t, s) Y(s) \, ds$ and $\hat{X}(t) \in T$ then Corollary 7-2.1 is true.

It $t = t_1$, the engineer says that $A(t, s)$ defines a *linear filter*. We shall discuss this case first. In what follows we develop a differential equation that is a necessary condition for $A(t, s)$ to satisfy in order to be optimal in the sense of minimum variance whenever $R(r)$ in (7-5b) is positive definite.

By differentiating (7-10) with respect to t_1 and interchanging $\partial/\partial t_1$ and the expectation operator E, one can obtain for all $s \in [t_0, t_1)$:

$$\frac{\partial}{\partial t_1}[\text{Cov}\, X(t_1) Y(s)] = \frac{\partial}{\partial t_1} E[X(t_1) Y^T(s)]$$

$$= E\left\{ \frac{\partial}{\partial t_1}[X(t_1)] Y^T(s) \right\} \tag{7-13}$$

Sec. 7-2 The Dynamic Linear Model

$$= E[\{F(t_1)X(t_1) + G(t_1)u(t)\}Y^T(s)]$$

$$= F(t_1) \operatorname{Cov}[X(t_1), Y(s)] + G(t_1) \operatorname{Cov}[U(t_1)Y(s)]$$

Since $U(t_1)$ is uncorrelated with $V(S)$ and $X(S)$ when $s < t_1$, then

$$\frac{\partial}{\partial t_1} \operatorname{Cov}[X(t_1), Y(s)] = F(t_1) \operatorname{Cov}[X(t_1), Y(s)]. \tag{7-14}$$

Since the $\operatorname{Cov}[Y(s), Y(w)] = \operatorname{Cov}[H(s) + V(s), H(w)X(w) + V(w)]$, it follows that

$$\frac{\partial}{\partial t_1} \int_{t_0}^{t_1} A(t_1, s) \operatorname{Cov}[Y(s), Y(w)] \, ds$$

$$= \frac{\partial}{\partial t_1} \left\{ \int_{t_0}^{t_1} A(t_1, s) \operatorname{Cov}[H(s)X(s), H(w)X(w)] \, ds \right.$$

$$\left. + \int_{t_0}^{t_1} A(t_1, s) \operatorname{Cov}[V(s), V(w)] \, ds \right\}$$

$$= A(t_1, t_1) \operatorname{Cov}[H(t_1)X(t_1), H(w)X(w)]$$

$$+ \int_{t_0}^{t_1} \frac{\partial}{\partial t_1} A(t_1, s) \operatorname{Cov}[H(s)X(s), H(w)X(w)] \, ds$$

$$+ \int_{t_0}^{t_1} \frac{\partial}{\partial t_1} A(t_1, s) \operatorname{Cov}[V(s), V(w)] \, ds$$

$$= A(t_1, t_1) \operatorname{Cov}[H(t_1)X(t_1), H(w)X(w)]$$

$$+ \int_{t_0}^{t_1} \frac{\partial}{\partial t_1} A(t_1, s) \operatorname{Cov}[Y(s), Y(w)] \, ds. \tag{7-15}$$

Note that

$$\operatorname{Cov}[H(t_1)X(t_1), H(w)X(w)] = \operatorname{Cov}[H(t_1)X(t_1), Y(w) - V(w)]$$

$$= H(t_1) \operatorname{Cov}[X(t_1)Y(w)] - \operatorname{Cov}[H(t_1)X(t_1), V(w)]$$

$$= H(t_1) \operatorname{Cov}[X(t_1), Y(w)]$$

since for all $w < t_1$, $\operatorname{Cov}[H(t_1)X(t_1), V(w)] = \phi$. Thus it follows that

$$\operatorname{Cov}[H(t_1)X(t_1), H(w)X(w)] = H(t_1) \operatorname{Cov}[X(t_1), Y(w)]. \tag{7-16}$$

On combining (7-13)–(7.16) we obtain

$$F(t_1) \operatorname{Cov}[X(t_1), Y(w)] - \int_{t_0}^{t_1} \frac{\partial}{\partial t_1} A(t_1, s) \operatorname{Cov}[Y(s), Y(w)] \, ds$$

$$- A(t_1, t_1) H(t_1) \operatorname{Cov}[X(t), Y(w)] = 0,$$

which implies

$$[F(t_1) - A(t_1, t_1) H(t_1)] \operatorname{Cov}[X(t_1), Y(w)]$$

$$- \int_{t_0}^{t_1} \frac{\partial}{\partial t_1} A(t_1, s) \operatorname{Cov}[Y(s), Y(w)] \, ds = 0.$$

But by 7-10,

$$\operatorname{Cov}[X(t_1), Y(w)] = \int_{t_0}^{t_1} A(t_1, s) \operatorname{Cov}[Y(s), Y(w)] \, ds,$$

which implies that

$$\int_{t_0}^{t_1} [F(t_1) A(t_1, s) - \frac{\partial}{\partial t_1} A(t_1, s) \qquad\qquad (7\text{-}17)$$

$$- A(t_1, t_1) H(t_1) A(t_1, s)] \operatorname{Cov}[Y(s), Y(w)] \, ds = 0$$

for all w such that $t_0 \leq w < t_1$.

Clearly, if $A(t_1, s)$ is a solution of

$$F(t_1) A(t_1, s) - \frac{\partial}{\partial t_1} A(t_1, s) - A(t_1, t_1) G(t_1) A(t_1, s) = 0 \qquad (7\text{-}18)$$

for every $t_0 \leq s \leq t_1$, then $A(t_1, s)$ is a solution of (7-17).

THEOREM 7-2.2. *Let* $R(s)$ *be positive definite in the interval* $t_0 \leq s \leq t_1$. *Then (7-18) is a necessary condition if* $A(t_1, s)$ *is to satisfy the Wiener–Hopf equation (7-10).*

Proof: If $A(t_1, s)$ satisfies the Wiener–Hopf equation, then

$$\hat{X}(t_1) = \int_{t_0}^{t_1} A(t_1, s) Y(s) \, ds$$

is an optimal estimator for $X(t_1)$. Let

$$B(t_1, s) = F(t_1) A(t_1, s) - \frac{\partial}{\partial t_1} A(t_1, s) - A(t_1, t_1) H(t_1) A(t_1, s).$$

Sec. 7-2 The Dynamic Linear Model

Then by (7-17),

$$\int_{t_0}^{t_1} B(t_1, s) \operatorname{Cov}[Y(s), Y(w)]\, ds = 0,$$

which implies that $A(t_1, s) + B(t_1, s)$ satisfies (7-10). Thus

$$\hat{X}(t_1) + \int_{t_0}^{t_1} B(t_1, s) Y(s)\, ds$$

is an optimal estimator for $X(t_1)$. By applying Lemma 7-2.1 we have

$$\left\| \hat{X}(t_1) + \int_{t_0}^{t_1} B(t_1, s) Y(s)\, ds - \hat{X}(t_1) \right\| = 0,$$

which implies

$$\left\| \int_{t_0}^{t_1} B(t_1, s) Y(s)\, ds \right\| = 0$$

or

$$\int_{t_0}^{t_1} \int_{t_0}^{t_1} B(t_1, s) \operatorname{Cov}[Y(s), Y(s')] B^T(t_1, s')(ds\, ds') = 0. \qquad (7\text{-}19)$$

From (7-4) it follows that

$$\operatorname{Cov}[Y(s), Y(s')] = \operatorname{Cov}[H(s)X(s), H(s')X(s')] + \operatorname{Cov}[V(s), V(s')]$$

$$= \operatorname{Cov}[H(s)X(s), H(s')X(s')] + R(s)\delta(s - s').$$

Thus it follows that (7-19) reduces to

$$\int_{t_0}^{t_1} \int_{t_0}^{t_1} (B(t_1, s)\{\operatorname{Cov}[H(s)X(s), H(s')X(s')] \\ + R(s)\delta(s - s')\} B^T(t_1, s'))\, ds\, ds' = 0. \qquad (7\text{-}20)$$

But $\operatorname{Cov}[H(s)X(s), H(s')X(s')]$ is nonnegative definite and $R(s)$ is positive definite for $t_0 \leq s \leq t_1$, which implies that (7-20) cannot be satisfied unless $B(t_0, s) = 0$ for every $t_0 \leq s \leq t_1$. Thus it follows that (7-18) is a necessary condition.

THEOREM 7-2.3. *The optimal estimate $\hat{X}(t_1)$ and the error $\tilde{X}(t_1) = X(t_1) - \hat{X}(t_1)$ satisfy*

$$\frac{d\hat{X}(t_1)}{dt_1} = F(t_1)\hat{X}(t_1) + A(t_1, t_1)[Y(t_1) - H(t_1)\hat{X}(t_1)]$$

and

$$\frac{d\tilde{X}(t_1)}{dt_1} = [F(t_1) - A(t_1, t_1)H(t_1)]\tilde{X}(t_1) + G(t_1)U(t_1) \\ - A(t_1, t_1)V(t_1), \quad (7\text{-}21)$$

respectively.

Proof: If we differentiate (7-6) with respect to t_1 we obtain

$$\frac{d\hat{X}(t_1)}{dt_1} = A(t_1, t_1)Y(t_1) + \int_{t_1}^{t_1} \frac{\partial}{\partial t_1} A(t_1, s) Y(s)\, ds.$$

From Theorem 7-2.2 it follows that

$$\frac{\partial A(t_1, s)}{\partial t_1} = F(t_1)A(t_1, s) - A(t_1, t_1)H(t_1)A(t_1, s),$$

which implies

$$\frac{d\hat{X}(t_1)}{dt_1} = K(t_1)Y(t_1) + F(t_1)\hat{X}(t_1) - A(t_1, t_1)H(t_1)\hat{X}(t_1)$$

$$= F(t_1)\hat{X}(t_1) + A(t_1, t_1)[Y(t_1) - H(t_1)\hat{X}(t_1)].$$

Similarly it can be shown that

$$\frac{d\tilde{X}(t_1)}{dt_1} = [F(t_1) - A(t_1, t_1)H(t_1)]\tilde{X}(t_1) + G(t_1)U(t_1) \\ - A(t_1, t_1)V(t_1).$$

LEMMA 7-2.2. *Let $A(t_1, s)$ satisfy (7-11). Then for every $t_0 \leq w \leq t_1$,*

$$\text{Cov}\,[X(t_1), H(w)X(w)] - \int_{t_1}^{t_1} A(t_1, s)\, \text{Cov}\,[H(s)(Xs), H(w)X(w)]\, ds \\ = A(t_1, w)R(w). \quad (7\text{-}22)$$

Proof: Let $t_0 \leq w < t_1$; then

$\text{Cov}\,[\tilde{X}(t_1), Y(w)]$

$$= \text{Cov}\left[X(t_1) - \int_{t_0}^{t_1} A(t_1, s) Y(s)\, ds,\ Y(w)\right]$$

$$= \text{Cov}\,[X(t_1), Y(w)] - \int_{t_0}^{t_1} A(t_1, s)\, \text{Cov}\,[Y(s), Y(w)]\, ds$$

Sec. 7-2 The Dynamic Linear Model

$$= \text{Cov}\,[X(t_1),\ Y(w)] - \int_{t_0}^{t_1} A(t_1,\,s)\,\text{Cov}\,[H(s)X(s),\ H(w)X(w)]\,ds$$

$$- \int_{t_0}^{t_1} A(t_1,\,s)R(s)\delta(s-w)\,ds$$

$$= \text{Cov}\,[X(t_1),\ Y(w)] - \int_{t_0}^{t_1} A(t_1,\,s)\,\text{Cov}\,[H(s)X(s),\ H(w)X(w)]\,ds$$

$$- A(t_1,\,w)R(w) = 0.$$

But since $Y(w) = H(w)X(w) + V(w)$ and $X(t_1)$ and $V(w)$ are independent, then

$$\text{Cov}\,[X(t_1),\ Y(w)] = \text{Cov}\,[X(t_1),\ H(w)X(w)],$$

which implies

$$\text{Cov}\,[X(t_1),\ H(w)X(w)] - \int_{t_0}^{t_1} A(t_1,\,s)\,\text{Cov}\,[H(s)X(s),\ H(w)X(w)]\,ds$$

$$= A(t_1,\,w)R(w).$$

Since both sides of the above equation are continuous functions in w, then it can be shown that the equation holds for $w = t_1$. Therefore, (7-22) holds for $t_0 \leq w \leq t_1$.

THEOREM 7-2.4. Let $\text{Cov}\,[\tilde{X}(t_1),\ \tilde{X}(t_1)] = P(t_1)$. Then

$$A(t_1,\,t_1) = P(t_1)H^T(t_1)R^{-1}(t_1).$$

Proof:

$$\text{Cov}\,[\tilde{X}(t_1),\ H(t_1)X(t_1)] = \text{Cov}\,[X(t_1) - \int_{t_0}^{t_1} A(t_1,\,s)Y(s)\,ds,\ H(t_1)X(t_1)]$$

$$= \text{Cov}\,[X(t_1),\ H(t_1)X(t_1)] - \text{Cov}\,[\int_{t_0}^{t_1} A(t_1,\,s)H(s)X(s)\,ds,\ H(t_1)X(t_1)]$$

$$- \text{Cov}\,\left[\int_{t_0}^{t_1} A(t_1,\,s)V(s)\,ds,\ H(t_1)X(t_1)\right]$$

$$= \text{Cov}\,[X(t_1),\ H(t_1)X(t_1)] - \int_{t_0}^{t_1} A(t_1,\,s)\,\text{Cov}\,[H(s)X(s),\ H(t_1)X(t_1)]\,ds$$

since

$$\int_{t_0}^{t_1} A(t_1,\,s)\,\text{Cov}\,[V(s),\ H(t_1)X(t_1)]\,ds = 0.$$

Thus
$$\text{Cov}[\tilde{X}(t_1), H(t_1)X(t_1)] = \text{Cov}[X(t_1), H(t_1)X(t_1)]$$
$$- \int_{t_0}^{t_1} A(t_1, s)\,\text{Cov}[H(s)X(s), H(t_1)X(t_1)]\,ds$$

and if we apply Lemma 7-2.2. with $w = t_1$, then
$$\text{Cov}[\tilde{X}(t_1), H(t_1)X(t_1)] = A(t_1, t_1)R(t_1).$$

But
$$\text{Cov}[\tilde{X}(t_1), H(t_1)X(t_1)] = \text{Cov}[\tilde{X}(t_1), X(t_1)]H^T(t_1),$$

which implies
$$\text{Cov}[\tilde{X}(t_1), X(t_1)]H^T(t_1) = A(t_1, t_1)R(t_1).$$

By Corollary 7-2.1,
$$\text{Cov}[\tilde{X}(t_1), \hat{X}(t_1)] = 0,$$

which implies
$$\text{Cov}[\tilde{X}(t_1), \tilde{X}(t_1)] = \text{Cov}[\tilde{X}(t_1), X(t_1) - \hat{X}(t_1)]$$
$$= \text{Cov}[\tilde{X}(t_1), X(t_1)].$$

Hence it follows that
$$\text{Cov}[\hat{X}(t_1), X(t_1)]H^T(t_1) = \text{Cov}[\tilde{X}(t_1), \tilde{X}(t_1)]H^T(t_1)$$
$$= P(T_1)H^T(t_1).$$

Therefore,
$$A(t_1, t_1)R(t_1) = P(t_1)H^T(t_1),$$

which implies
$$A(t_1, t_1) = P(t_1)H^T(t_1)R^{-1}(t_1).$$

It is well known that the general solution of (7-7) is of the form
$$X(t) = \Phi(t, t_0)X(t_0) + \int_{t_0}^{t} \Phi(t, s)G(s)U(s)\,ds, \tag{7-23}$$

where Φ is the transition or fundamental matrix that satisfies the following relations:
$$\frac{d\Phi(t, t_0)}{dt} = F(t)\Phi(t, t_0),$$

Sec. 7-2 The Dynamic Linear Model

$$\Phi(t_0, t_0) = I \text{ (identity matrix)} \quad \text{for all } t_0,$$

$$\Phi^{-1}(t_0, t_1) = \Phi(t_1, t_0) \quad \text{for every } t_0, t_1,$$

and

$$\Phi(t_1, t_0) = \Phi(t_1, t)\Phi(t, t_0) \quad \text{for every } t_0, t, \text{ and } t_1.$$

THEOREM 7-2.5. *The variance* $P(t_1)$ *satisfies*

$$\frac{dP(t_1)}{dt_1} = F(t_1)P(t_1) + P(t_1)F^T(t_1) - P(t_1)H^T(t_1)R^{-1}(t_1)H(t_1)P(t_1) \\ + G(t_1)Q(t_1)G^T(t_1). \qquad (7\text{-}24)$$

Proof:

$$\frac{dP(t_1)}{dt_1} = \text{Cov}\left[\frac{d\tilde{X}(t)}{dt_1}, \tilde{X}(t_1)\right] + \text{Cov}\left[\tilde{X}(t_1), \frac{d\tilde{X}(t_1)}{dt_1}\right]$$

and by applying Theorem 7-2.3,

$$\frac{dP(t_1)}{dt_1} = \text{Cov}\{[F(t_1) - A(t_1, t_1)H(t_1)]\tilde{X}(t_1), \tilde{X}(t_1)\} \\ + \text{Cov}[G(t_1)U(t_1), \tilde{X}(t_1)] + \text{Cov}[-A(t_1, t_1)V(t_1), \tilde{X}(t_1)] \\ + \text{Cov}[\tilde{X}(t_1), [F(t_1) - A(t_1, t_1)H(t_1)]\tilde{X}(t_1)] \\ + \text{Cov}[\tilde{X}(t_1), G(t_1)U(t_1)] + \text{Cov}[\tilde{X}(t_1), -A(t_1, t_1)V(t_1)].$$

Now let $\Phi(t_1, s)$ be the common transition matrix for the two differential equations of Theorem 7-2.3; then

$$\text{Cov}\,[\tilde{X}(t_1), U(t_1)] = \text{Cov}\left[\int_{t_0}^{t} \Phi(t_1, s)[G(s)U(s) - A(s, s)V(s)]\,ds,\, U^T(t_1)\right]$$

since

$$\tilde{X}(t_1) = \Phi(t_1, t_0)\tilde{X}(t_0) + \int_{t_0}^{t_1} \Phi(t_1, s)[G(s)U(s) - A(s, s)V(s)]\,ds.$$

Thus

$$\text{Cov}\,[\tilde{X}(t_1), U(t_1)] = \int_{t_0}^{t_1} \Phi(t_1, s)G(s)Q(s)\delta(s - t_1)\,ds$$

$$= \tfrac{1}{2}\Phi(t_1, t_1)G(t_1)Q(t_1),$$

$$= \tfrac{1}{2}G(t_1)Q(t_1).$$

Similarly it can be shown that

$$\text{Cov}\,[\tilde{X}(t_1),\,V(t_1)] = -\tfrac{1}{2}K(t_1)R(t_1).$$

Therefore,

$$\frac{dP(t_1)}{dt_1} = [F(t_1) - A(t_1, t_1)H(t_1)]P(t_1) + \tfrac{1}{2}G(t_1)Q(t_1)G^T(t_1)$$

$$+ \tfrac{1}{2}A(t_1, t_1)R(t_1)A^T(t_1, t_1) + P(t_1)[F^T(t_1) - H^T(t_1)K^T(t_1)]$$

$$+ \tfrac{1}{2}G(t_1)Q(t_1)G^T(t_1) + \tfrac{1}{2}A(t_1, t_1)R(t_1)A^T(t_1, t_1).$$

Using the fact that

$$A(t_1, t_1) = P(t_1)H^T(t_1)R^{-1}(t_1),$$

then

$$\frac{dP(t_1)}{dt_1} = F(t_1)P(t_1) + P(t_1)F^T(t_1) - P(t_1)H^T(t_1)R^{-1}(t_1)H(t_1)P(t_1)$$

$$+ G(t_1)Q(t_1)G^T(t_1).$$

To consider the linear prediction problem, let $t > t_1$; then by (7-23),

$$X(t) = \Phi(t, t_1)X(t_1) + \int_{t_1}^t \Phi(t_1, s)G(s)U(s)\,ds.$$

Now in the interval $t_1 < s < t$ $U(s)$ is independent of $X(s)$ in the interval $t_0 \leq s \leq t_1$, which implies that the optimal linear estimate of

$$\int_{t_0}^t \Phi(t_1, s)G(s)U(s)\,ds$$

is 0. Hence it follows that

$$\hat{X}(t) = \Phi(t, t_1)\hat{X}(t_1). \qquad (7\text{-}25)$$

Also it follows by (7-6) and Theorem 7-2.1 that

$$\hat{X}(t_0) = 0.$$

Thus

$$P(t_0) = \text{Cov}\,[\tilde{X}(t_0),\,\tilde{X}(t_0)]$$

$$= \text{Cov}\,[X(t_0),\,X(t_0)].$$

7-3 Concluding Remarks

The solution of the equations

$$\frac{d\hat{X}(t_1)}{dt} = F(t)\hat{X}(t_1) + A(t_1, t_1)[Y(t_1) - H(t_1)\hat{X}(t_1)],$$

$$\frac{dP(t_1)}{dt_1} = F(t_1)P(t_1) + P(t_1)F^T(t_1) - P(t_1)H(t_1)^T R^{-1}(t_1)H(t_1)P(t_1)$$
$$+ G(t_1)Q(t_1)G^T(t_1),$$

$$A(t_1, t_1) = P(t_1)H^T(t_1)R^{-1}(t_1), \quad \text{and}$$

$$\hat{X}(t_0) = \phi$$

yields the otpimal estimator for $X(t_1)$. This estimator is known to many engineers as the Kalman or Kalman–Bucy filter. [97, 100].

EXERCISES

1. Show that

$$\frac{d\tilde{X}(t_1)}{dt_1} = [F(t_1) - A(t_1, t_1)H(t_1)]\tilde{X}(t_1)$$
$$+ G(t_1)U(t_1) - A(t_1, t_1)V(t_1).$$

2. Since both sides of the equation

$$\text{Cov}\,[X(t_1), H(w)X(w)] - \int_{t_0}^{t_1} A(t_1, s)\,\text{Cov}\,[H(s)X(s), H(w)X(w)]\,ds$$
$$= A(t_1, w)R(w)$$

are continuous functions in w, then show that the equation holds for $w = t_1$.

3. Show that

$$\text{Cov}\,[\hat{X}(t_1|t_1), V(t_1)] = -\tfrac{1}{2}K(t_1)R(t_1).$$

CHAPTER **8**

Concerning Best Estimators Using an Estimated Covariance Matrix

8-1 Introduction

In this chapter we consider some problems in estimating the unknown non-random parameter x in the linear model

$$Y = Hx + U, \qquad (8\text{-}1)$$

as defined previously in Theorem 3-3.1. If V, the covariance matrix of U, is known, then the Gauss–Markov theorem applies and the best estimator of x is given by Equation (3-9), that is,

$$\hat{X} = (H^T V^{-1} H^T)^{-1} H^T V^{-1} Y. \qquad (8\text{-}2)$$

We consider here the case when V is unknown, but an estimate $\hat{V}$ of V is available. This estimate may be computed from previous data, or from the present data making no assumptions concerning x.

One unbiased estimator that is immediately available, but in general not minimum variance but only minimal least square is

$$X_{LS} = (H^T H)^{-1} H^T Y, \qquad (8\text{-}3)$$

which has proved to be undesirable in many cases, especially if V is "much different" from $\sigma^2 I$. Clearly, if V can be assumed to be $\sigma^2 I$, where the scalar σ^2 is unknown, (8-3) is minimum variance.

It is noted here, as shown by Rao [160], that the estimator of x, obtained by merely substituting an estimator $\hat{V}$ for V in (8-2) is not necessarily best; that is, it may be possible to use known or inferred knowledge of the covariance matrix V to obtain an estimator with better characteristics than the estimator obtained by substituting a known $\hat{V}$ in (8-2).

8-2 Pertinent Matrix Identities

To facilitate the reading of what comes later and to make the chapter reasonably self-contained Lemmas 8-2.1 through 8-2.3 are listed and proved.

LEMMA 8-2.1. *Let* A *be a positive definite matrix partitioned as*

$$A = \begin{bmatrix} A_{11} & A_{12} \\ A_{21} & A_{22} \end{bmatrix} \quad (8\text{-}4)$$

with its inverse as

$$A^{-1} = \begin{bmatrix} B_{11} & B_{12} \\ B_{21} & B_{22} \end{bmatrix};$$

then $B_{11} - (A_{11})^{-1}$ *is nonnegative definite.*

Proof: From Theorem 1-4.9, we know that

$$B_{11} = A_{11}^{-1} + A_{11}^{-1} A_{12} [A_{22} - A_{21} A_{11}^{-1} A_{12}]^{-1} A_{21} A_{11}^{-1}.$$

Hence

$$B_{11} - A_{11}^{-1} = A_{11}^{-1} A_{12} [A_{22} - A_{21} A_{11}^{-1} A_{12}]^{-1} A_{21} A_{11}^{-1},$$

a nonnegative definite matrix.

LEMMA 8-2.2. *Let* H *be an* Np $\times$ n *matrix of rank* n, *and let* Z *be an* Np $\times$ (Np $-$ n) *matrix of rank* (Np $-$ n) *such that* $H^T Z = \phi$. *Then*

$$(H^T V^{-1} H)^{-1} H^T V^{-1} = (H^T H)^{-1} H^T - (H^T H)^{-1} H^T V Z (Z^T V Z)^{-1} Z^T, \quad (8\text{-}5)$$

where V *is any* Np $\times$ Np *positive definite matrix.*

Proof: Let

$$L_1 = (H^T V^{-1} H)^{-1} H^T V^{-1}, \quad \text{and}$$

$$L_2 = (H^T H)^{-1} H^T - (H^T H)^{-1} H^T V Z (Z^T V Z)^{-1} Z^T,$$

where $H^T Z = \phi$, rank $(H) = n$, and rank $(Z) = Np - n$. Let $\{Y\}$ denote the Euclidean Np space and let y be any element from $\{Y\}$. Then

$$y = \sum_{i=1}^{n} a_i H^i + \sum_{j=1}^{Np-n} b_j Z^j,$$

where H^i $(i = 1, 2, \ldots, n)$ and Z^j $(j = 1, 2, \ldots, Np - n)$ are the linearly independent column vectors of H and Z, respectively. We must show that $L_1 y = L_2 y$ for every $y \in \{Y\}$, hence $L_1 = L_2$. This is equivalent to showing that $L_1 H = L_2 H$ and $L_1 Z = L_2 Z$.

Multiplying both sides of (8-5) by H, it is computational to verify that $L_1 H = L_2 H$, since

$$L_1 H = (H^T V^{-1} H) H^T V^{-1} H = I$$

and

$$L_2 H = [(H^T H)^{-1} H^T - (H^T H)^{-1} H^T V Z (Z^T V Z)^{-1} Z^T] H = I$$

the $n \times n$ identity matrix.

We note that

$$P_1 = Z(Z^T Z)^{-1} Z^T$$

and

$$P_2 = H(H^T H)^{-1} H^T$$

are idempotent such that

$$\text{rank } (P_1) = Np - n$$

$$\text{rank } (P_2) = n$$

and

$$P_1 P_2 = \phi,$$

which implies that

$$I = P_1 + P_2.$$

It follows that

$$\phi = H^T Z$$

$$= H^T V^{-1} I V Z$$

$$= H^T V^{-1} [Z(Z^T Z)^{-1} Z^T + H(H^T H)^{-1} H^T] V Z$$

$$= H^T V^{-1} Z[(Z^T Z)^{-1} Z^T V Z] + H^T V^{-1} H(H^T H)^{-1} H^T V Z.$$

This implies that

$$(H^TV^{-1}Z)[(Z^TZ)^{-1}Z^TVZ] = -H^TV^{-1}H(H^TH)^{-1}H^TVZ$$

or

$$H^TV^{-1}Z = -H^TV^{-1}H(H^TH)^{-1}H^TVZ[(Z^TZ)^{-1}Z^TVZ]^{-1},$$

and finally,

$$(H^TV^{-1}H)^{-1}H^TV^{-1}Z = (H^TH)^{-1}H^TZ - (H^TH)^{-1}H^TVZ[(Z^TZ)^{-1}Z^TVZ]^{-1}$$

since

$$(H^TH)^{-1}H^TZ = \phi;$$

that is, $L_1Z = L_2Z$. Lemma 8-2.2 is proved.

LEMMA 8-2.3. *With* H *an* Z *as in Lemma* 8-2.2, *the matrix*

$$(H^TH)^{-1}H^TVH(H^TH)^{-1} - (H^TV^{-1}H^T)^{-1} \quad (8\text{-}6)$$

is nonnegative definite.

Proof: Consider the matrix

$$\begin{bmatrix} H^TV^{-1}H & H^TV^{-1}Z \\ Z^TV^{-1}H & Z^TV^{-1}Z \end{bmatrix} = \begin{bmatrix} H^T \\ Z^T \end{bmatrix} V^{-1}[H : Z]. \quad (8\text{-}7)$$

Since [H : Z] is nonsingular we know that the inverse of the right-hand side is

$$[H : Z]^{-1} V \begin{bmatrix} H^T \\ Z^T \end{bmatrix}^{-1}, \quad (8\text{-}8)$$

where

$$[H : Z]^{-1} = \begin{bmatrix} (H^TH)^{-1}H^T \\ (Z^TZ)^{-1}Z^T \end{bmatrix}. \quad (8\text{-}9)$$

Substituting (8-9) into (8-8) and computing, one can obtain (8-8) in a form with the upper left-hand partition given by

$$(H^TH)^{-1}H^TVH(H^TH)^{-1}.$$

But by Lemma 8-2.1 we know that (8-6) holds.

8-3 Covariance Adjustment

Let T_1 and T_2 be two vector estimators with dimensions $n \times 1$ and $r \times 1$, respectively. Also, let

$$E(T_1) = x$$

and

$$E(T_2) = \phi.$$

The vector T_1 is an unbiased estimator of x, but if

$$R_{12} = \text{Cov}(T_1, T_2) \neq \phi.$$

then an estimator of x can be found with smaller or at least equal variance whenever the covariance matrix

$$R = \begin{bmatrix} R_{11} & R_{12} \\ R_{21} & R_{22} \end{bmatrix} \qquad (8\text{-}10)$$

of (T_1, T_2) is known.
    Consider the estimator

$$X^* = T_1 - R_{12} R_{22}^{-1} T_2.$$

Then the covariance matrix V^* of X^* is

$$V^* = R_{11} + R_{12} R_{22}^{-1} R_{22} R_{22}^{-1} R_{21}$$
$$- 2 R_{12} R_{22}^{-1} R_{21}$$
$$= R_{11} - R_{12} R_{22}^{-1} R_{21}$$

or

$$R_{11} - V^* \qquad (8\text{-}11)$$

is at least positive semidefinite. Hence X^* is a better estimator or at least as good as T_1.
    However, if only an estimator of R

$$\hat{R} = \begin{bmatrix} \hat{R}_{11} & \hat{R}_{12} \\ \hat{R}_{21} & \hat{R}_{22} \end{bmatrix}$$

is available, we substitute $\hat{R}_{ij}$ for R_{ij} in Equation (8-10) and obtain what Rao [160] calls the (covariance) adjusted estimator

$$\tilde{X} = T_1 - \hat{R}_{12}\hat{R}_{22}^{-1}T_2. \tag{8-12}$$

Now when $\hat{R}_{ij}$, $i = 1, 2, j = 1, 2$ are independently distributed with T_i, $i = 1, 2$, $\tilde{X}$ is an unbiased estimator for x. Also, under these same conditions, the covariance matrix for $\tilde{X}$, say, $\tilde{V}$, follows directly from (8-12) and is given by

$$\tilde{V} = R_{11} + E[\hat{R}_{12}\hat{R}_{22}^{-1}R_{22}\hat{R}_{22}^{-1}\hat{R}_{21}] - E[\hat{R}_{12}\hat{R}_{22}\hat{R}_{21}] - E[\hat{R}_{12}\hat{R}_{22}^{-1}\hat{R}_{21}],$$

$$\tag{8-13}$$

where the expectations are taken with respect to the density of $\hat{R}_{12}\hat{R}_{22}^{-1}$. Note that

$$R_{11} - \tilde{V}$$

may not necessarily be nonnegative definite as was

$$R_{11} - V^*$$

in (8-11). Note that if R_{12} is equal to or "close" to ϕ, then $\tilde{X}$ is a poorer estimator of x than is T_1; that is, $R_{11} - \tilde{V}$ is negative definite.

Caution is necessary in the technique of covariance adjustment since it can result in a descrease in efficiency when the covariance matrix is used in place of the unknown matrix. However, if R_{12} is not close to ϕ, one should expect $\tilde{V} - R_{11}$ to be negative definite and $\tilde{X}$ to be better than T_1.

In conclusion, then, it is possible that the use of T_2 may not be optimal. In the absence of the exact knowledge of R, the optimal estimator cannot be found. However, an estimate of R may lead to some choice of T_2 that may lead to better estimators.

8-4 An Arbitrary Covariance Matrix V

If V is known, then the best linear unbiased estimator for x is

$$\hat{X} = (H^T V^{-1} H)^{-1} H^T V^{-1} Y$$

and a natural estimator for x when V is not known but $\hat{V}$ an estimate of V is known is given by

$$X^* = (H^T \hat{V}^{-1} H)^{-1} H^T \hat{V}^{-1} Y.$$

Sec. 8-4 An Arbitrary Covariance Matrix V

However, let

$$T_1 = (H^T H)^{-1} H^T Y, \qquad E(T_1) = x \qquad (8\text{-}14)$$

and

$$T_2 = Z^T Y, \qquad E(T_2) = \phi, \qquad (8\text{-}15)$$

where Z is the $Np \times (Np - n)$ matrix of rank $(Np - n)$ such that $Z^T H = \phi$.

Since V is not known, one must estimate R as defined by (8-10) by the estimator $\hat{R}$, where in this case

$$\hat{R} = \begin{bmatrix} (H^T H)^{-1} H^T \hat{V}^{-1} H^T (H^T H)^{-1} & (H^T H)^{-1} H^T \hat{V}^{-1} Z \\ Z^T \hat{V}^{-1} (H^T H)^{-1} & Z^T \hat{V}^{-1} Z \end{bmatrix}.$$

On judicious application of the lemma in Equation (8-3) the adjusted estimator follows from Equation (8-12); that is,

$$T_1 - \hat{R}_{12} \hat{R}_{22}^{-1} T_2 = (H^T H)^{-1} H^T Y - (H^T H)^{-1} H^T \hat{V} Z (Z^T \hat{V} Z)^{-1} Z^T Y$$

$$= (H^T \hat{V}^{-1} H)^{-1} H^T \hat{V}^{-1} Y. \qquad (8\text{-}16)$$

Thus the estimators $T_1 - \hat{R}_{12} \hat{R}_{22}^{-1} T_2$ and X^* are the same. Note that this result is true for our selection of T_1 and T_2. There may be other estimators T_1 and T_2 that may increase the efficiency.

We have noted that X_{LS} given by (8-3) is best when $V = \sigma^2 I$. A pertinent question that can be answered is the following: What class of covariance matrices V exist such that

$$(H^T V^{-1} H)^{-1} H^T V^{-1} = (H^T H)^{-1} H^T ? \qquad (8\text{-}17)$$

Theorem 8-4.1 provides this class.

THEOREM 8-4.1. *Let Z be an* $Np \times Np - n$ *matrix of rank* $Np - n = Np - \text{rank}(H)$ *such that* $Z^T H = \phi$, *and let* $\mathscr{V}$ *be the set of matrices of the form*

$$V = H V_1 H^T + Z V_2 Z^T + \sigma^2 I, \qquad (8\text{-}18)$$

where V_1, V_2 *and* σ^2 *(a scalar) are arbitrary. Then the necessary and sufficient condition that the least squares estimator of* x, *in the model* $Y = Hx + U$ *with* $\text{Cov}(U) = V$, *be the same as that for the special choice* $\text{Cov}(U) = \sigma^2 I$ *is that* $V \in \mathscr{V}$.

Proof: The complete class of linear functions of Y with zero expectation is given by $Z^T Y$. Hence, if $X_{LS} = (H^T H)^{-1} H^T Y$ is the least square estimator for x, then by (8-11) for $R_{12} = 0$,

$$\text{Cov}\,[X_{LS}, Z^T Y] = \text{Cov}\,[(H^T H)^{-1} H^T Y, Z Y]$$

$$= (H^T H)^{-1} H^T (V + H X X^T H^T) Z$$

$$= (H^T H)^{-1} H^T V Z.$$

That is, $(H^T H)^{-1} H^T V Z = 0$. A necessary and sufficient condition for this to be true is $H^T V Z = 0$. It is computational to verify that $H^T V Z = \phi$ is true if and only if V is of the form defined by (8-18).

Also, one notes that there may exist conditions on the covariance matrix V such that the ordinary least square estimator (8-3) is equal to the minimum variance estimator (8-2). Theorem 8-4.2 summarizes these conditions.

THEOREM 8-4.2. Necessary and sufficient conditions for (8-3) and (8-2) to be the same estimators is that the covariance matrix **V** *in (8-2) must be of the form*

$$(1 - p)\mathbf{I} + p\mathbf{J}\mathbf{J}^T,$$

where p is a scalar such that

$$0 \le p < 1$$

and

$$\mathbf{J}^T = (1, 1, \ldots, 1).$$

The proof is left to the reader as an exercise and can be found in Reference [123].

Theorem 8-4.2 indicates that in practice it will be unlikely for (8-2) to be the same as (8-3).

8-5 Concluding Remarks

It is clear that the problem of obtaining optimal estimators for x when V is not known is indeed a complicated one. Born [29] has written a recursive estimator when V is not known but assumed block diagonal with equal diagonal blocks. He achieved satisfactory results in his application to an orbit determination problem.

EXERCISES

1. Let H and Z be defined as in Lemma 8-2.2. Show that $Z(Z^TZ)^{-1}Z^T$ and $H(H^TH)^{-1}H^T$ are idempotent such that rank $[Z(Z^TZ)^{-1}Z^T] = Np - n$ and rank $[H(H^TH)^{-1}H^T] = n$. Also, show that $I = Z(Z^TZ)^{-1}Z^T + H(H^TH)^{-1}H^T$.

2. Prove Theorem 8-4.1 when the rank of H is less than full rank.

3. Show that if
$$V = HV_1H + ZV_2Z^T + V_0$$
in Theorem 8-4.1, then the least squares estimator of x with Cov $(U) = V$ is the same as if Cov $(U) = V_0$.

4. Let $Y = H_1x + H_2T + U$ such that $E(T) = \phi$, $E(U) = \phi$, Var $(U) = \Delta^2 I$, Var $(T) = R$, and Cov $(T, U) = \phi$. Also, let H_1 and H_2 be known matrices. Suppose further that there exists a matrix C such that $H_1^T(H_2 - H_1C) = \phi$ and that the rank of $(H_2 - H_1C)$ is n. Considering the linear transformations,
$$T_1 = (H_1^TH_1)^{-1}H_1^TY,$$
$$T_2 = (H_2 - H_1C)^TY,$$
and
$$T_3 = A^TY,$$
where $A^TH_1 = \phi$ and $A^T(H_2 - H_1C)^T = \phi$. Show that $E(T_3) = \phi$, Cov $(T_1, T_3) = \phi$, and Cov $(T_2, T_3) = \phi$. Does the estimator T_3 have smaller variance than T_1 or T_2?

5. Prove Theorem 8-4.2.

CHAPTER 9

On Selecting an Optimal Design Matrix

9-1 Introduction

In this chapter we consider the problem of estimating the coefficients in a full-rank linear model

$$Y = Hx + u. \tag{9-1}$$

We let Y be an $N \times 1$ observable random vector, H an $N \times (p+1)$ design matrix of rank $p+1$ whose rows are the points at which observations are made, x a $(p+1) \times 1$ parameter vector to be estimated, and U an $N \times 1$ random error vector such that

$$E(U) = 0$$

$$E(UU^T) = V,$$

where V is a known positive definite $N \times N$ matrix.

It is well known (Gauss–Markov theorem) that the best linear unbiased estimator $\hat{X}$ of x is given by

$$\hat{X} = (H^T V^{-1} H)^{-1} H^T V^{-1} Y$$

with variance

$$\text{Var}(\hat{X}) = (H^T V^{-1} H)^{-1}.$$

Note that the variance of $\hat{X}$ is a function of the design matrix H. It is desirable, then, to select the design matrix H in such a manner as to minimize the diagonal elements of the matrix $(H^T V^{-1} H)^{-1}$. Thus the corresponding estimator $\hat{X}$ has been improved in the sense of minimizing the variances of the elements of $\hat{X}$.

To define what we mean by an optimum design matrix H so that the diagonal elements of $(H^T V^{-1} H)^{-1}$ are minimized we make the following definition. In the discussion to follow, we shall make $V = \sigma^2 I$ since there will be no loss in generality.

Definition 9-1.1. *A matrix* H *is said to be as good as a matrix* Z, *written* H $\leq$ Z, *if* $(Z^T Z)^{-1} - (H^T H)^{-1}$ *is positive semidefinite.*

Definition 9-1.2. *A matrix* H *is said to be better than a matrix* Z, *written* H < Z, *if* $(Z^T Z)^{-1} - (H^T H)^{-1}$ *is positive semidefinite and not null.*

Definition 9-1.3. *A matrix* H *is said to be equivalent to a matrix* Z, *written* H $\simeq$ Z, *if* $(H^T H)^{-1} = (Z^T Z)^{-1}$.

LEMMA 9-1.1. *The relation* $\leq$ *defined above satisfies the following two properties of a partial ordering:*

1. *For any matrix* H, H $\leq$ H.
2. *If* H $\leq$ Z *and* Z $\leq$ W, *then* H $\leq$ W.

Proof: To show property 1, note that for any matrix H, $(H^T H)^{-1} - (H^T H)^{-1} = 0$, which is positive semidefinite. Thus $H \leq H$ for every matrix H.

To show property 2, we observe that if $H \leq Z$ and $Z \leq W$, then both $(Z^T Z)^{-1} - (H^T H)$ and $(W^T W)^{-1} - (Z^T Z)^{-1}$ are positive semidefinite. Hence the sum $(W^T W)^{-1} - (Z^T Z)^{-1} + (Z^T Z)^{-1} - (H^T H)$ is positive semidefinite, which implies $H \leq W$.

Example 9-1.1 shows why the antisymmetric property, $H \leq Z$ and $Z \leq H$ implies $H = Z$, of partial orderings does not hold here. It should be noted, however, that the two matrices in Example 9-1.1 are somewhat related in that they are both "optimal" designs for a special case discussed in Excercises 4, 5, and 7 at the end of this chapter.

Example 9-1.1. Let

$$H = \begin{bmatrix} 1 & 1 & 0 \\ 1 & -1 & 0 \\ 1 & 0 & 1 \\ 1 & 0 & -1 \end{bmatrix} \quad \text{and} \quad Z = \begin{bmatrix} 1 & \frac{1}{\sqrt{2}} & \frac{1}{\sqrt{2}} \\ 1 & \frac{1}{\sqrt{2}} & -\frac{1}{\sqrt{2}} \\ 1 & -\frac{1}{\sqrt{2}} & -\frac{1}{\sqrt{2}} \\ 1 & -\frac{1}{\sqrt{2}} & \frac{1}{\sqrt{2}} \end{bmatrix}.$$

Then it follows that

$$(H^T H)^{-1} = (Z^T Z)^{-1} = \begin{bmatrix} \frac{1}{4} & 0 & 0 \\ 0 & \frac{1}{2} & 0 \\ 0 & 0 & \frac{1}{2} \end{bmatrix}.$$

The standard definitions for comparability for partial ordering follow.

Definition 9-1.4. *Two matrices* H *and* Z *are said to be comparable if either* H $\leq$ Z *or* Z $\leq$ H.

It is easy to show that all matrices are not comparable; for suppose

$$(H^T H)^{-1} = \begin{bmatrix} 2 & 0 & 0 \\ 0 & 3 & 0 \\ 0 & 0 & 1 \end{bmatrix}$$

and

$$(Z^T Z)^{-1} = \begin{bmatrix} 1 & 0 & 0 \\ 0 & 4 & 0 \\ 0 & 0 & 1 \end{bmatrix}.$$

It then follows that

$$(H^T H)^{-1} - (Z^T Z)^{-1}$$

or

$$(Z^T Z)^{-1} - (H^T H)^{-1}$$

are not positive semidefinite. Since all matrices are not comparable under Definition 9-1.3 and 9-1.4, we add Definitions 9-1.5 and 9-1.6.

Definition 9-1.5. *The matrix* H *is said to be weakly better than the matrix* Z, *written* H $\underset{w}{\leq}$ Z, *if*

$$\max \{a_{ii}\} \leq \max \{b_{ii}\},$$

where

$$(H^T H)^{-1} = (a_{ij}) \quad \text{and} \quad (Z^T Z)^{-1} = (b_{ij}).$$

Definition 9-1.6. *The matrix* H *is said to be weakly better than the matrix* Z, *written* H $\underset{w}{<}$ Z, *if*

$$\max \{a_{ii}\} < \max \{b_{ii}\}.$$

LEMMA 9-1.2. The following properties hold:

1. *For any matrix* H, *then* $H \underset{w}{\leq} H$.
2. *If* $H \underset{w}{\leq} Z$ *and* $Z \underset{w}{\leq} W$, *then* $H \underset{w}{\leq} W$.
3. *For any two matrices, then either* $H \underset{w}{\leq} Z$ *or* $Z \underset{w}{\leq} H$.

Proof: To prove property 1, let H be any matrix, then $(H^TH)^{-1} = (H^TH)^{-1}$, which implies max $\{a_{ii}\}$ = max $\{a_{ii}\}$. Thus $H \underset{w}{\leq} H$.

To prove property 2, let $H \underset{w}{\leq} Z$ and $Z \underset{w}{\leq} H$; then max $\{a_{ii}\} \leq$ max $\{b_{ii}\}$ and max $\{b_{ii}\} \leq$ max $\{c_{ii}\}$, where $(H^TH)^{-1} = (a_{ij})$, $(Z^TZ)^{-1} = (b_{ij})$, and $(W^TW) = (c_{ij})$. Hence max $\{a_{ij}\} \leq$ max $\{b_{ii}\} \leq$ max $\{c_{ii}\}$ implies max $\{a_{ii}\} \leq$ max $\{c_{ii}\}$ and consequently, $H \underset{w}{\leq} W$.

To prove property 3, let H and Z be any two matrices, where $(H^TH)^{-1} = (a_{ij})$ and $(Z^TZ)^{-1} = (b_{ij})$. Let $a =$ max $\{a_{ii}\}$ and $b =$ max $\{b_{ii}\}$; then since a and b are real numbers we have either $a \leq b$ or $b \leq a$. Therefore, $H \underset{w}{\leq} Z$ or $Z \underset{w}{\leq} H$.

Notice that the relation $\underset{w}{\leq}$ satisfies all of the properties for a total ordering except for the antisymmetric property defined above. However, we do have the property that any two matrices are comparable under this definition.

The relationship between the inequalities $\leq$ and $\underset{w}{\leq}$ is demonstrated by Lemma 9-1.3.

LEMMA 9-1.3. If $H \leq Z$, *then* $H \underset{w}{\leq} Z$.

Proof: Let $(H^TH)^{-1} = (a_{ij})$ and $(Z^TZ)^{-1} = (b_{ij})$. Then $H \leq Z$ implies that $(Z^TZ)^{-1} - (H^TH)^{-1}$ is positive semidefinite or that $(b_{ij}) - (a_{ij})$ is positive semidefinite. But each diagonal element $b_{ii} - a_{ii}$ is a principal submatrix of $(Z^TZ)^{-1} - (H^TH)^{-1}$ and has a nonnegative determinant. Therefore, $b_{ii} - a_{ii} \leq 0$ or $a_{ii} \leq b_{ii}$ for all i. Thus max $\{a_{ii}\} \leq$ max $\{b_{ii}\}$ and $H \underset{w}{\leq} Z$.

The following additional properties of the relation $\leq$ will help to demonstrate our interest in it with respect to the present problem of minimizing Var $(\hat{\beta})$ by selecting the design matrix H. The first of these properties could have been chosen as the basic definition and since matrix inversion in practice is at best tedious or inexact, Property 9-1.1. will be used extensively.

PROPERTY 9-1.1. $H \leq Z$ *if and only if* $H^TH - Z^TZ$ *is positive semidefinite.*

Sec. 9-1   Introduction   149

Proof: Let $A_1 = (H^T H)$, $A_2 = (Z^T Z)$, and $A = A_1 - A_2$. Suppose A is positive semidefinite. Then $A_1 = A_2 + A$ or $A_1^{-1} = (A_2 + A)^{-1} = (A_2 + C^T C)^{-1} = A_2^{-1} - A_2^{-1} C^T (I + C A_2^{-1} C^T)^{-1} C A_2^{-1}$, where $A = C^T C$. Therefore, $A_2^{-1} - A_1^{-1} = A_2^{-1} C^T (I + C A_2^{-1} C^T)^{-1} C A_2^{-1}$ is positive semidefinite since A_2^{-1} and $C^T (I + C A_2^{-1} C^T)^{-1} C$ are positive semidefinite. Thus $(Z^T Z)^{-1} - (H^T H)^{-1}$ is positive semidefinite and hence $H \leq Z$.

The converse can be similarly shown.

PROPERTY 9-1.2. *If* $H = (J, H_1, \ldots, H_p)$, $Z = (J, Z_1, \ldots, Z_p)$, *and* $T = (\phi, t_1 J, \ldots, t_p J)$, *where the t_i's are known constants, then* $H \leq Z$ *implies that* $(H + T) \leq (Z + T)$.

Proof: From the definition of H, Z, and T we see that $(H + T) = (J, H_1 + t_1 J, \ldots, H_p + t_p J)$ and $(Z + T) = (J, Z_1 + t_1 J, \ldots, Z_p + t_p J)$. Therefore, by multiplication we have

$(H + T)^T (H + T) =$

$$= \begin{bmatrix} J^T J & J^T (H_1 + t_1 J) & \cdots & J^T (H_p + t_p J) \\ (H_1 + t_1 J)^T J & (H_1 + t_1 J)^T (H_1 + t_1 J) & \cdots & (H_1 + t_1 J)^T (H_p + t_p J) \\ \vdots & \vdots & & \vdots \\ (H_p + t_p J)^T J & (H_p + t_p J)^T (H_1 + t_1 J) & \cdots & (H_p + t_p J)^T (H_p + t_p J) \end{bmatrix}$$

$$= \begin{bmatrix} J^T J & J^T H_1 + t_1 J^T J & \cdots & J^T H_p + t_p J^T J \\ H_1^T J + t_1 J^T J & H_1^T H_1 + t_1 H_1^T J \\ & + t_1 J^T H_1 + t_1 t_1 J^T J & \cdots & H_1^T H_p + t_p H_1^T J \\ & & & + t_1 J^T H_p + t_1 t_p J^T J \\ \vdots & \vdots & & \vdots \\ H_p^T J + t_p J^T J & H_p^T H_1 + t_1 H_p^T J \\ & + t_p J^T H_1 + t_p t_1 J^T J & \cdots & H_p^T H_p + t_p H_p^T J \\ & & & + t_p J^T H_p + t_p t_p J^T J \end{bmatrix}$$

Consider the nonsingular matrix of elementary operations:

$$E = \begin{bmatrix} 1 & -t_1 & -t_2 & \cdots & -t_p \\ 0 & 1 & 0 & \cdots & 0 \\ \vdots & \vdots & \vdots & & \vdots \\ 0 & 0 & 0 & \cdots & 1 \end{bmatrix}.$$

Then by multiplication, $E^T(H+T)^T(H+T)E =$

$$\begin{bmatrix} J^TJ & J^TH_1 & \cdots & J^TH_p \\ H_1^TJ & H_1^TH_1 & \cdots & H_1^TH_p \\ \cdot & \cdot & & \cdot \\ \cdot & \cdot & & \cdot \\ \cdot & \cdot & & \cdot \\ H_p^TJ & H_p^TH_1 & \cdots & H_p^TH_p \end{bmatrix},$$

which is H^TH. That is, $(H+T)^T(H+T) = E^{-T}H^THE^{-1}$, where $E^{-T} = (E^{-1})^T$.

Similarly, we see that $E^T(Z+T)^T(Z+T)E = Z^TZ$ or $(Z+T)^T(Z+T) = E^{-T}Z^TZE^{-1}$. But by Property 9-1.1, $H \leq Z$ implies that $H^TH - Z^TZ$ is positive semidefinite. The difference $(H+T)^T(H+T) - (Z+T)^T(Z+T) = E^{-T}H^TE^{-1} - E^{-T}Z^TZE^{-1} = E^{-T}(H^TH - Z^TZ)E^{-1}$ is positive semidefinite since $H^TH - Z^TZ$ is positive semidefinite. Hence, by Property 9-1.1, $(H+T) \leq (Z+T)$.

PROPERTY 9-1.3. *If* R *is a nonsingular matrix, then* $H \leq Z$ *implies that* $HR \leq ZR$.

Proof: $(HR)^T(HR) - (ZR)^T(ZR) = R^TH^THR - R^TZ^TZR = R^T(H^TH - Z^TZ)R$. But $H^TH - Z^TZ$ is positive semidefinite since $H \leq Z$. Thus $R^T(H^TH - Z^TZ)R$ is positve semidefinite and $HR \leq ZR$.

PROPERTY 9-1.4. *If* R *is an orthogonal matrix, then* $H \leq Z$ *implies that* $RH \leq RZ$.

Proof: $(RH)^T(RH) - (RZ)^T(RZ) = H^TR^TRH - Z^TR^TRZ = H^TIH - Z^TIZ = H^TH - Z^TZ$. But $H \leq Z$; hence $H^TH - Z^TZ$ is positive semidefinite and $(RH)^T(RH) - (RZ)^T(RZ)$ is positive semidefinite and, therefore, $RH \leq RZ$.

PROPERTY 9-1.5. *If* $H \leq Z$, *then* $|H^TH| \geq |Z^TZ|$.

Proof: Let $A_1 = H^TH$ and $A_2 = Z^TZ$. Then by Property 9-1.1 we have $H \leq Z$ implying that $A = A_1 - A_2 = H^TH - Z^TZ$ is positive semidefinite. Hence $A_1 = A + A_2$ and $|A_1| = |A + A_2| \geq |A| + |A_2| \geq |A_2|$ by Minkowski's determinant theorem. Thus $|H^TH| \geq |Z^TZ|$.

PROPERTY 9-1.6. *If* $H \leq Z$, *then* tr $(H^TH) \leq$ tr (Z^TZ).

Proof: Let $H^TH = (a_{ij})$ and $Z^TZ = (b_{ij})$. Then $H \leq Z$ implies by Property 9-1.1 that $H^TH - Z^TZ$ is positive semidefinite. Therefore, each diagonal element $a_{ii} - b_{ii}$ is a principal submatrix of $H^TH - Z^TZ$, and has a nonnegative determinant. Thus $a_{ii} - b_{ii} \geq 0$ for all i and hence tr $(H^T H)$ = $\sum_{i=1}^{p+1} a_{ii} - \sum_{i=1}^{p+1} b_{ii} =$ tr $(Z^T Z)$.

9-2 Rectangular and Ellipsoidal Regions

We now consider Var $(\hat{X}) = \sigma^2 (H^T H)^{-1}$ to see if there are any natural restrictions placed on the selection of H. Theorem 9-2.1 by Rao [157] is of interest.

THEOREM 9-2.1. *Let* Var $(X) = \sigma^2 (H^T H)^{-1}$ *be the covariance matrix of the best linear unbiased estimate for* x *in the linear model* Y $=$ Hx $+$ u. *Then if one can select the columns of* H $= (H_0, H_1, \ldots, H_p)$ *such that the scalar* $H_i^T H_i$ *is maximal and* $H_i^T H_j = 0$ *for all* i $\neq$ j, *the* Var $(\hat{X})$ *is minimal.*

Proof: Let

$$\frac{\text{Var}(\hat{X})}{\sigma^2} = \begin{bmatrix} V_{11} & V_{12} \\ V_{21} & V_{22} \end{bmatrix} = \begin{bmatrix} R_{11} & R_{12} \\ R_{21} & R_{22} \end{bmatrix}^{-1}.$$

Then $V_{11} = (R_{11} - R_{12} R_{22}^{-1} R_{21})^{-1} = R_{11}^{-1} + R_{11}^{-1} R_{12} V_{22} R_{21} R_{11}^{-1}$. To minimize V_{11} we certainly want to choose R_{12} such that $R_{11}^{-1} R_{12} V_{22} R_{21} R_{11}^{-1} = 0$. But since $(H^T H)^{-1}$ is positive definite, then V_{22} is positive definite. Hence $R_{21} R_{11}^{-1} = R_{11}^{-1} R_{12} = 0$ or $R_{12} = R_{21} = 0$. But by definition R_{12} is a matrix whose elements are of the form $H_i^T H_j$ for $i \neq j$. Thus $H_i^T H_j = 0$ for all $i \neq j$. If R_{11} is a scalar then $R_{11} = H_1^T H_1$ and consequently $v_{11} = (H_1^T H_1)^{-1}$. Thus, to minimize v_{11} we must maximize $H_1^T H_1$, by selecting H_1 as far away from the $N \times 1$ zero vector as possible in the l_2 norm sense. This will be true for any of the diagonal elements v_{ii} of V since multiplication on the left by the elementary matrix E interchanges rows 1 and i and on the left by E^T interchanges v_{ii} and v_{11}. Hence v_{ii} is minimized when $H_i^T H_i$ is maximized, that is, when H_i is selected as far from the $N \times 1$ zero vector as possible.

Theorem 9-2.1 indicates that one should select H_i at the $N \times 1$ "infinity" vector, then Var $(\hat{X}) = 1/\sigma^2 (v_{ij}) = 0$. But this is not practical for two reasons.

    1. The linearity assumptions on which the model is based are valid in only a bounded region R in E_{p+1}.

and/or

    2. The physical limitations on sampling restrict our attention to a bounded region R in E_{p+1}.

Thus $R = \{(h_0, h_1, \ldots, h_p): \sum_{i=0}^{p} h_i^2 \leq c\}$, for some positive real number c.

We shall consider two special regions, rectangular and ellipsoidal, given by

$$R_1 = \{1, h_1, \ldots, h_p\} : (a_1, \ldots, a_p) \leq (h_1, \ldots, h_p)$$
$$\leq (b_1, \ldots, b_p)\}$$

and

$$R_2 = \{(1, h_1, \ldots, h_p) : (h_1, \ldots, h_p) A (h_1, \ldots, h_p)^T$$
$$\leq 1,$$

where A is a $p \times p$ positive definite matrix}.

Using Theorems 9-2.2 through 9-2.4 we see that we shall need to consider only those rectangular and ellipsoidal regions whose centers are at the origin and whose axes are parallel to the coordinate axes.

Definition 9-2.1. *A translation matrix* T *is an* $N \times p + 1$ *matrix of the form* $T = (\phi, t_1 J, \ldots, t_p J)$, *where* ϕ *is the* $N \times 1$ *vector of zeros,* J *is the* $N \times 1$ *vector of ones, and the constants* t_i *are known.*

It should be noted that for the regions described above the design matrix H for the *regression* linear model [9-1] can be written in the form

$$H = (J, H_1, \ldots, H_p).$$

THEOREM 9-2.2. (Translation). *Let* H *and* Z *be any two design matrices of the form above and let* T *be a translation matrix. Then* $H + T \leq Z + T$ *if and only if* $H \leq Z$.

Proof: It is clear that the design and translation described here are precisely the matrices of Property 9-1.2; hence Theorem 9-2.2 is proved.

THEOREM 9-2.3. *If* H *is a design matrix and* T *is a translation matrix then* $\det((H + T)^T(H + T)) = \det(H^T H)$.

Proof: In Property 9-1.2 we described a matrix E such that $(H + T)^T(H + T) = E^{-T} H^T H E^{-1}$. But E is a product of elementary column operations matrices each of which has determinant 1. Thus E has determinant 1, and $\det[(H + T)^T(H + T)] = \det(E^{-T} H^T H E^{-1}) = \det(E^{-T}) \cdot \det(H^T H) \det(E^{-1}) = [\det(E^{-1})]^2 \det(H^T H) = \det(H^T H)$.

Definition 9-2.2. *A rotation matrix is a* $p + 1 \times p + 1$ *orthogonal matrix.*

THEOREM 9-2.4. (Rotation) *Let* H *and* Z *be any two design matrices*

Sec. 9-2   Rectangular and Ellipsoidal Regions

of the form above and let R be a rotation matrix. Then $HR \leq ZR$ if and only if $H \leq Z$.

Proof: Since R is orthogonal, $R^T = R^{-1}$ and by Property 9-1.3, $HR \leq ZR$ if $H \leq Z$.

THEOREM 9-2.5. Let R be a $p+1 \times p+1$ nonsingular matrix; then for any design matrix H $det\,[(HR)^T(HR)] = det\,(R)^2\,det\,(H^TH)$.

Proof: $det\,[(HR)^T(HR)] = det\,(R^TH^THR) = det\,(R^T)\,det\,(H^TH)\,det\,(R) = det\,(R)\,det\,(R^T)\,det\,(H^TH) = det\,(RR^T)\,det\,(H^TH) = [det\,(R)]^2\,det\,(H^TH)$.

COROLLARY 9-2.1. If H is a design matrix and R is a rotation matrix then $det\,[(HR)^T(HR)] = det\,(H^TH)$.

Proof: In Theorem 9-2.5 we have that $det\,[(HR)^T(HR)] = det\,(RR^T)\,det\,(H^TH)$. But $RR^T = I$ and thus $det\,(RR^T) = 1$. Hence $det\,[(HR)^T(HR)] = det\,(H^TH)$.

Definition 9-2.3. A scaling matrix is a $p+1 \times p+1$ diagonal matrix with all diagonal elements positive constants.

THEOREM 9-2.6. Let H and Z be any two design matrices and let S be a scaling matrix. Then $HS \leq ZS$ if and only if $H \leq Z$.

Proof: Since S is a diagonal matrix with all diagonal elements positive constants, then S is nonsingular. Thus H and S are the matrices of Property 9-1.3 and $HS \leq ZS$ if $H \leq Z$.

Since the relation $\leq$ is less than a partial ordering and not all designs are comparable, there may not be an optimal design, so we define a minimal class of designs from which a best design may be selected.

Definition 9-2.4. A design H is said to be admissible if there is no design Z better than H.

Definition 9-2.5. A class of admissible designs A is said to be complete if for every design Z not in A there is a design H in A that is better.

Definition 9-2.6. A class of admissible designs A is said to be minimal complete if A has no complete subsets.

Definition 9-2.7. A class of admissible designs A is said to be essentially complete if for every design H not in A there is a design in A that is at least as good.

Definition 9-2.8. *A class of admissible designs* A *is said to be minimally essentially complete if* A *has no essentially complete subsets.*

The problem now is to classify those regions R that have an optimal design and those regions that have a nontrivial minimal complete class of designs, and to describe the optimal design or the minimal complete class of designs.

EXERCISES

1. Suppose the parameter vector x is constrained such that $Rx = r$, where R is an $N \times n$ known matrix and r is an $N \times 1$ known vector along with the constraint (9-1). Find the minimum mean-square error estimator of x.

2. Let x be an $n \times 1$ vector and R be any $N \times n$ known matrix such that

$$(xx^T)^{-1} - (xx^T)^{-1}R^T[R(xx^T)^{-1}R^T]^{-1}R(xx^T)^{-1}$$

 exists. Show that to minimize the diagonal elements of the above expression one need only minimize the diagonal elements of $(xx^T)^{-1}$.

3. Suppose the parameter vector x is constrained such that $a \leq Rx \leq b$, where R is an $N \times n$ known matrix and a and b are known $N \times 1$ vectors along with the constraint (9-1). Find the minimum mean-square error estimator of x.

4. Let H be an $N \times (p+1)$ design matrix. Let H_i be the ith column of H and constrain H such that $H_i^T V^{-1} H_i \leq C_i$, $i = 1, \ldots, p+1$, where C_i, $i = 1, \ldots, p+1$ and V^{-1} are known. Let $\hat{X}_i$ be the minimum variance linear unbiased estimator of x_i. Then show that Var $(\hat{X}_i) \geq (1/C_i)$ and the minimum variance is realized when $(H_i^T V^{-1} H_j) = 0$ for $i \neq j$ and $(H_i^T V^{-1} H_i) = C_i$.

5. Let a random sample of size N (even) be taken, where $Y_i = x_1 + x_2 h_i + U_i$, $i = 1, 2, \ldots, N$ or matrix form $Y = Hx + U$, where

$$H = \begin{bmatrix} 1 & h_1 \\ 1 & h_2 \\ \cdot & \cdot \\ \cdot & \cdot \\ \cdot & \cdot \\ 1 & h_N \end{bmatrix}, \quad x = \begin{bmatrix} x_1 \\ x_2 \end{bmatrix}$$

and

$$U = \begin{bmatrix} U_1 \\ U_2 \\ \cdot \\ \cdot \\ \cdot \\ U_N \end{bmatrix}.$$

Furthermore, assume $E(U) = 0$, $E(UU^T) = \sigma^2 I$ and that $h_i \in [-a, a]$, where a is a real number. To minimize $\text{Var}(\hat{X}_1)$ and $\text{Var}(\hat{X}_2)$ determine where the h_i's should be chosen.

6. Let N be odd in Exercise 5.

7. Given the linear model $Y = Hx + U$ with sample size n (n divisible by 4) such that

   1. $Y_i = x_1 + x_2 h_{i1} + x_3 h_{i2} + U_i$.
   2. $E(U) = 0$ and $E(UU^T) = \sigma^2 I$.
   3. $(h_{i1}, h_{i2}) \in C$ for every $i = 1, 2, \ldots, n$, where C is a rectangle with vertices (a, b), $(-a, -b)$, $(a, -b)$, and $(-a, b)$.

   Find the best choices of (h_{i1}, h_{i2}) so that $\text{Var}(\hat{X}_j)$, $j = 1, 2, 3$ is minimized for each j.

8. Consider Exercise 7, where

   1. $n - 1 = 4k$
   2. $n - 2 = 4k$
   3. $n - 3 = 4k$, where k is a positive integer.

9. Let C in Exercise 7 be an ellipse centered at the origin with end points of the major axis at $(-a, 0)$ and $(a, 0)$ and end points of the minor axis as $(0, -b)$ and $(0, b)$.

CHAPTER 10

Related Topics in Parametric Estimation

10-1 Introductory Remarks

If the scientist or engineer *knows* the form of the density function of the random errors, it is possible in some cases to compute the minimum variance unbiased estimator for the parameter vector x instead of the minimum variance *linear* unbiased estimator for x. However, in the case of normality assumptions the best linear unbiased estimator for x is also the best unbiased estimator. This is shown again in this chapter in a different way using the so-called Cramer–Rao lower bound for the variance matrix.

Also, we introduce in this chapter the concept of a sufficient statistic, which is important in parametric statistic, since the concept leads to a theory and methodology that can lead one to minimum variance unbiased estimators of various parameters of interest.

10-2 The Matrix Lower Bound

A matrix lower bound for the covariance matrices of vector estimators can be formulated in a matrix notation that facilitates the search for vector estimators with minimal covariance matrices. (We recall that one positive definite covariance matrix is by definition less than or equal to another provided the second minus the first is nonnegative definite.) The derivation presented here is similar to that of Cramer in which he established the efficiency of vector estimators using the concept of ellipsoids of concentration.

A matrix lower bound for the covariance matrices of unbiased vector estimators of the unknown parameters in the linear regression model with *correlated* normal error is established. A direct result of this application is that the best linear estimator given by the Gauss–Markov theorem still yields the minimum covariance matrix when compared with other vector estimators from the larger class composed of linear and nonlinear vector estimators. This result has been noted for the case of uncorrelated normal errors.

Let the joint density function of n random variables $Y = [Y_1, \ldots, Y_n]^T$ be

$$L = L(Y_1, \ldots, Y_n; \beta_1, \ldots, \beta_p), \tag{10-1}$$

where $\beta = [\beta_1, \ldots, \beta_p]^T$ are unknown parameters that we wish to estimate. An arbitrary $n \times m$ matrix A will at times be denoted by $\{a_{ij}\}_n^m$.

By definition of joint density function, it follows that

$$1 = \int_{-\infty}^{\infty} \cdots \int_{-\infty}^{\infty} L \, dy_1 \cdots dy_n. \tag{10-2}$$

It is assumed that the following regularity conditions hold:

1. $\dfrac{\partial}{\partial \beta_i} \left[\int_{-\infty}^{\infty} \cdots \int_{-\infty}^{\infty} L \, dy_1 \cdots dy_n \right] = \int_{-\infty}^{\infty} \cdots \int_{-\infty}^{\infty} \dfrac{\partial L}{\partial \beta_i} dy_1 \cdots dy_n,$

   $i = 1, 2, \ldots, p.$

2. $\dfrac{\partial}{\partial \beta_j} \left[\int_{-\infty}^{\infty} \cdots \int_{-\infty}^{\infty} t_i L \, dy_1 \cdots dy_n \right] = \int_{-\infty}^{\infty} \cdots \int_{-\infty}^{\infty} t_i \dfrac{\partial L}{\partial \beta_j} dy_1, \ldots, dy_n,$

   $i, j = 1, 2, \ldots, p,$

where $T = [T_1, T_2, \ldots, T_p]^T$ is a vector function of the elements of Y. The vector T is an unbiased vector estimator for β.

On differentiating both sides of (10-2) with respect to β_i and by condition (1), one obtains

$$0 = \int_{-\infty}^{\infty} \cdots \int_{-\infty}^{\infty} \frac{\partial L}{\partial \beta_i} dy_1 \cdots dy_n$$

$$0 = \int_{-\infty}^{\infty} \cdots \int_{-\infty}^{\infty} \left(\frac{1}{L} \frac{\partial L}{\partial \beta_i} L \, dy_1 \cdots dy_n \right) \tag{10-3}$$

$$0 = E\left[\frac{\partial \ln L}{\partial \beta_i} \right] \quad i = 1, 2, \ldots, p. \tag{10-4}$$

Sec. 10-2 The Matrix Lower Bound

Let the $p \times 1$ vector S of random variables be defined by

$$S = \left\{\frac{\partial \ln L}{\partial \beta_i}\right\}, \qquad i = 1, \ldots, p \tag{10-5}$$

with mean vector

$$E(S) = \Phi, \tag{10-6}$$

and a $p \times p$ covariance matrix $E(SS^T) = \Lambda$, that is,

$$\Lambda = \left\{E\left[\frac{\partial \ln L}{\partial \beta_i}\frac{\partial \ln L}{\partial \beta_j}\right]\right\}, \qquad \begin{matrix}i = 1, \ldots, p \\ j = 1, \ldots, p\end{matrix} \tag{10-7}$$

a positive definite matrix.

Let T be an unbiased vector estimator of β; then

$$\beta_i = \int_{-\infty}^{\infty} \cdots \int_{-\infty}^{\infty} t_i L \, dy_1 \cdots dy_n, \qquad i = 1, 2, \ldots, n. \tag{10-8}$$

On differentiating with respect to β_j both sides of (10-8), it follows, if condition 2 holds, that

$$\delta_{ij} = \int_{-\infty}^{\infty} \cdots \int_{-\infty}^{\infty} t_i \frac{\partial L}{\partial \beta_j} \, dy_1 \cdots dy_n = \int_{-\infty}^{\infty} \cdots \int_{-\infty}^{\infty} t_i \frac{\partial \ln L}{\partial \beta_j} l \, dy_1 \cdots dy_n \tag{10-9}$$

$$= E[T_i S_j],$$

where $\delta_{ij} = 1$, if $i = j$ and $\delta_{ij} = 0$, if $i \neq j$.

Let the covariance matrix of T be Σ; that is,

$$\Sigma = E[(T - \beta)(T - \beta)^T], \tag{10-10}$$

a positive definite matrix.

Consider the $2p \times 1$ vector V, which is built by adjoining the vectors T and S,

$$V = (T^T S^T)^T. \tag{10-11}$$

The covariance matrix R_V of V follows from (10-7), (10-9), and (10-10) and is

$$R_V = \begin{bmatrix} \Sigma & I \\ I & \Lambda \end{bmatrix} \tag{10-12}$$

since $E(TS^T) = I$ is the covariance matrix of the vectors T and S. We note that R_V is positive definite, which implies that R_V^{-1} is positive definite. Let

R_V^{-1} be partitioned compatibly with (10-11), that is,

$$R_V^{-1} = \begin{bmatrix} A_{11} & A_{12} \\ A_{21} & A_{22} \end{bmatrix},$$

where each submatrix A_{ij} is a $p \times p$ matrix. The inverse of the principal submatrix

$$A_{11}^{-1} = \Sigma - I\Lambda^{-1}I = \Sigma - \Lambda^{-1}$$

is also positive definite; hence we conclude that

$$\Sigma > \Lambda^{-1},$$

the desired matrix lower bound.

10-3 An Application

Let the $n \times 1$ observation vector Y be related to the $p \times 1$ vector of parameters according to the linear model

$$Y = X\beta + U, \qquad (10\text{-}14)$$

where X is a known $n \times p$ mapping matrix and U is an $n \times 1$ normal random vector, whose mean vector is $E(U) = \Phi$, and known $n \times n$ covariance matrix

$$E(UU^T) = R.$$

The probability density function of the observation vector Y is

$$L = \frac{1}{(2\pi)^{n/2}|R|^{1/2}} \exp\left\{-\frac{1}{2}[(y - X\beta)^T R^{-1}(y - X\beta)]\right\}. \qquad (10\text{-}15)$$

We note from (10-14) that

$$E(Y) = X\beta \qquad (10\text{-}16)$$

$$E(YY^T) = R + X\beta\beta^T X^T \qquad (10\text{-}17)$$

and from (10-15) that

$$\frac{\partial \ln L}{\partial \beta} = X^T R^{-1} Y - X^T R^{-1} X\beta. \qquad (10\text{-}18)$$

It follows from (10-7) and (10-18) that

$$\Lambda = E[(X^T R^{-1} Y - X^T R^{-1} X\beta)(X^T R^{-1} Y - X^T R^{-1} X\beta)^T]. \quad (10\text{-}19)$$

On expanding the right side of (10-19) and substituting (10-16) and (10-17), the inverse of the matrix bound is found to be

$$(X^T R^{-1} X). \quad (10\text{-}20)$$

The desired bound is by (10-13),

$$(X^T R^{-1} X)^{-1},$$

which is clearly the covariance matrix of the vector estimator given by the Gauss–Markov theorem, that is,

$$\hat{\beta} = (X^T R^{-1} X)^{-1} X^T R^{-1} Y.$$

It follows that the minimum variance unbiased vector estimate for $c^T \beta$, where c is a vector of constants, is simply $c^T \hat{\beta}$.

10-4 A Sufficient Statistic for a Parameter

Let X be an $n \times 1$ random vector with probability density function $f(X; \theta)$, where θ is a $k \times 1$ nonrandom parameter vector and an element of a given parameter space Ω. Let $X_1, X_2, \ldots, X_N$ be a random sample from the above distribution. The joint probability density function of $X_1, X_2, \ldots, X_N$ is given by

$$f(x_1; \theta) f(x_2; \theta) \cdots f(x_N; \theta).$$

Let $Y_1 = u_1(X_1, X_2, \ldots, X_N)$, $Y_2 = u_2(X_1, \ldots, X_N), \ldots, Y_N = u_N(X_1, X_2, \ldots, X_N)$ be any N statistic. If the transformation is one-to-one and if the random variables are continuous, then the joint density function is given by

$$h(y_1, y_2, \ldots, y_N; \theta) = f[w_1(y_1, \ldots, y_N; \theta)] f[w_2(y_1, \ldots, y_N; \theta)] \cdots$$
$$f[w_N(y_1, \ldots, y_N; \theta)] |J|,$$

where J is the Jacobian of the transformation; or if the random variables are discrete, then the probability density function is given by

$$h(y_1, y_2, \ldots, y_N; \theta) = f[w_1(y_1, \ldots, y_N; \theta)], \ldots, f[w_N(y_1, \ldots, y_N; \theta)].$$

In the case that the transformation is not one-to-one, the right-hand side is the sum of i terms of the same form, where i is the number of regions wherein the transformation is one-to-one. The conditional probability density function of $Y_1, Y_2, \ldots, Y_{i-1}, Y_{i+1}, \ldots, Y_N$ given $Y_i = y_i$ is

$$h(y_1, y_2, \ldots, y_{i-1}, y_{i+1}, \ldots, y_N | y_i; \theta) = \frac{h(y_1, \ldots, y_{i-1}, y_{i+1}, \ldots, y_N; \theta)}{h(y_i; \theta)},$$

(10-21)

where

$$h(y_1, \ldots, y_{i-1}, y_{i+1}, \ldots, y_N; \theta)$$

and

$$h(y_i; \theta) \tag{10-22}$$

are the joint probability density and marginal density functions of $Y_1, Y_2, \ldots, Y_{i-1}, Y_{i+1}, \ldots, Y_N$ and Y_i, respectively.

An important and desirable property for an estimator of θ to have is known as *sufficiency* of the estimator. In general the only information about θ in $f(x; \theta)$ is contained in the sample random variables $X_1, X_2, \ldots, X_N$. Therefore one desires an estimator, say, Y_i, that contains all the information about θ in the sample. When an estimator of θ has this property we call the estimator a sufficient statistic of θ. We now give a formal definition of sufficient statistic.

Definition 10-4.1. *Let $X_1, X_2, \ldots, X_N$ be a random sample from a distribution that is one member of the family $\{f(X; \theta), \theta \in \Omega\}$. Let $Y_1, Y_2, \ldots, Y_N$ be as defined above such that the Jacobian of the transformation is not identically zero. Then $\hat{\theta}$, say, $\hat{\theta} = Y_i$, is called a sufficient statistic of θ if and only if the conditional probability density function defined by (10-21) is independent of θ.*

To illustrate Definition 10-4.1 let us consider Example 10-4.1.

Example 10-4.1. Let X be a bivariate normal distribution such that

$$f_X(x; \mu) = \frac{1}{2\pi} e^{-1/2(x-\mu)^T(x-\mu)}, \quad -\infty < x < \infty.$$

Let X_1 and X_2 be a random sample from this bivariate population. Let $Y = X_1 + X_2$ and $Z = X_2$; then the Jacobian of this transformation is equal to 1. Now

$$f_{Y,Z}(y, z; \mu) = \frac{1}{(2\pi)^2} e^{-1/2[(1/2)(y-2\mu)^T(y-2\mu) + 2(z-y/2)^T(z-y/2)]},$$

$$-\infty < z < \infty, \quad z < y < \infty,$$

Sec. 10-4 A Sufficient Statistic for a Parameter

which implies

$$f_Y(y; \mu) = \frac{1}{4\pi} e^{-1/4(y-2\mu)^T(y-2\mu)}.$$

Hence

$$h(z \mid y; \mu) = \frac{f_{Y,Z}(y, z)}{f_Y(y)}$$

$$= \frac{1}{\pi} e^{-(z-y/2)^T(z-y/2)}, \quad -\infty < z < \infty,$$

which is independent of μ. Thus Y is a sufficient statistic of μ by definition.

In general it is quite a task to find a statistic that satisfies the definition for being a sufficient statistic. A criterion for aiding one to find a sufficient statistic is summarized in the following factorization theorems.

THEOREM 10-4.1. *Let $X_1, X_2, \ldots, X_N$ be a random sample from a distribution that is one member of the family $\{f(x; \theta), \theta \in \Omega\}$. Let $Y_1 = u_1(X_1, X_2, \ldots, X_N)$ be a statistic for the parameter θ. Then Y_1 is a sufficient statistic for θ if and only if there exists a function h such that $h(x_1, x_2, \ldots, x_N)$ does not depend on θ and*

$$\prod_{i=1}^{N} f(x_i; \theta) = g_1(u_1(x_1, \ldots, x_N); \theta) h(x_1, \ldots, x_N). \tag{10-23}$$

$g_1(u_1(x_1, \ldots, x_N); \theta)$ *is the density function for the random variable* $u_1(X_1, \ldots, X_N)$.

Proof: Define in addition to $y_1 = u_1(x_1, \ldots, x_N)$ the following functions: $y_2 = u_2(x_1, \ldots, x_N), \ldots, y_N = u_N(x_1, \ldots, x_N)$ having inverses $x_1 = w_1(y_1, \ldots, y_N), \ldots, x_N = w_N(y_1, \ldots, y_N)$. Thus if Y_1 is a sufficient statistic then

$$g(y_1, \ldots, y_N; \theta) = g_1(y_1 \mid \theta) t(y_2, \ldots, y_N \mid y_1; \theta),$$

where $t(y_2, \ldots, y_N \mid y_1; \theta)$ does not depend on θ. Since $g(y_1, \ldots, y_N; \theta)$ is the joint density function of $Y_1, \ldots, Y_N$, then

$$g[u_1(x_1, \ldots, x_N), \ldots, u_N(x_1, \ldots, x_N); \theta] |J|^{-1}$$

is the joint density function of $X_1, \ldots, X_N$, which implies

$$g[u_1(x_1, \ldots, x_N), \ldots, u_N(x_1, \ldots, x_N)] |J|^{-1} = \prod_{i=1}^{N} f(x_i; \theta).$$

Now $t[u_2(x_1, \ldots, x_N), \ldots, u_N(x_1, \ldots, x_N)] \mid y_1; \theta) |J|^{-1}$ does not depend on

θ so let us define $h = t|J|^{-1}$. Thus it follows that

$$\prod_{i=1}^{N} f(x_i; \theta) = g_1[u_1(x_1, \ldots, x_N); \theta] h(x_1, \ldots, x_N),$$

where h is a function not depending on θ.

To prove the converse suppose

$$\prod_{i=1}^{N} f(x_i; \theta) = g_1[u_1(x_1, \ldots, x_N); \theta] h(x_1, \ldots, x_N),$$

where $h(x_1, \ldots, x_N)$ does not depend on θ. Then

$$\prod_{i=1}^{N} f(x_i; \theta) |J| = g(y_1, \ldots, y_N; \theta),$$

where $g(y_1, \ldots, y_N; \theta)$ is the joint density for $Y_1, \ldots, Y_N$. Now $g_1(y_1; \theta)$ is the marginal density of Y_1 and $h[w_1(y_1, \ldots, y_N), \ldots, w_N(y_1, \ldots, y_N)]|J|$ is a function not depending on θ. In fact it follows that $h[w_1(y_1, \ldots, y_N), \ldots, w_N(y_1, \ldots, y_N)]|J|$ is the conditional density of $Y_2, \ldots, Y_N$ given $Y_1 = y_1$, which completes the proof.

It is obvious that in general it is a difficult task to find the density function of Y_1. Theorem 10-4.2 removes this difficulty.

THEOREM 10-4.2. *Let $X_1, \ldots, X_N$ be a random sample from a distribution that is one member of the family $\{f(x; \theta); \theta \in \Omega\}$. Let $Y_1 = u_1(X_1, \ldots, X_N)$ be a statistic for the parameter θ. Then Y_1 is a sufficient statistic for the parameter θ if and only if there exist functions h and k such that $h(x_1, \ldots, x_N)$ does not depend on θ and*

$$\prod_{i=1}^{N} f(x_i; \theta) = k[u_1(x_1, \ldots, x_N); \theta] h(x_1, \ldots, x_N).$$

The proof of Theorem 10-4.2 is similar to the proof of Theorem 10-3.1 and will be omitted.

Example 10-4.2. Let $X_1, X_2, \ldots, X_N$ be a random sample from multivariate normal $p \times 1$ random variable X such that $EX = \mu$ and Cov $X = I$. Show that $(1/N) \sum_{i=1}^{N} X_i$ is a sufficient statistic for μ.

Let $\bar{X}_N = (1/N) \sum_{i=1}^{N} X_i$. Then

$$f(x_1, x_2, \ldots, x_N; \mu, I) = \frac{1}{(2\pi)^{pN/2}} e^{-(1/2) \sum_{i=1}^{N} (x_i - \mu)(x_i - \mu)^T}$$

$$= \frac{1}{(2\pi)^{pN/2}} e^{-(1/2) \sum_{i=1}^{N} (x_i - \bar{x}_N)(x_i - \bar{x}_N)^T} e^{-(N/2)(\bar{x}_N - \mu)(\bar{x}_N - \mu)^T},$$

Sec. 10-4 A Sufficient Statistic for a Parameter

which implies that we can choose

$$k(\bar{x}_N; \mu) = \frac{1}{(2\pi)^{pN/2}} e^{-(N/2)(\bar{x}_N - \mu)(\bar{x}_N - \mu)^T}$$

and

$$h(x_1, \ldots, x_N) = e^{-(1/2)\sum_{i=1}^{N}(x_i - \bar{x}_N)(x_i - \bar{x}_N)^T}$$

Theorems 10-4.3 and 10-4.4 are of importance when we are searching for best estimators of the parameter θ, where best has been defined in Definition 2-4.4.

THEOREM 10-4.3. (Rao–Blackwell theorem). *Let Y be an* $n \times 1$ *random variable with finite mean μ and finite variance* V_Y. *Let X be an* $n \times 1$ *random variable and define* $E[Y|x] = g(x)$. *Then* $E[g(X)] = \mu$ *and* $V_Y - V_{g(X)}$ *is positive semidefinite.*

Proof: The proof will be given for X and Y being continuous random variables. If X and Y are discrete random variables, summations will replace the integrals and the proof is similar. Let $h(y|x)$ be the conditional density of Y given $X = x$ and $f(x)$, the density function of X. Thus

$$E[g(X)] = \left\{\int g(x_i)f(x)\,dx\right\}$$

$$= \left\{\iint y_i h(y_i|x)\,dy_i\,dx\right\}, \quad i = 1, 2, \ldots, n$$

$$= \mu.$$

To show that $V_Y - V_{g(X)}$ is positive semidefinite consider

$$V_Y = E[(Y - \mu)(Y - \mu)^T]$$

$$= E\{[Y - g(X) + g(X) - \mu][Y - g(X) + g(X) - \mu]^T\}$$

$$= E\{[Y - g(X)][Y - g(X)]^T\} + E\{[g(X) - \mu][g(X) - \mu]^T\}$$

$$+ E\{[g(X) - \mu][Y - g(X)]^T\} + E\{[Y - (gX)][g(X) - \mu]^T\}.$$

Now

$$E\{[Y - g(X)][g(X) - \mu]^T\} = E[Yg^T(X)] - E[g(X)g^T(X)].$$

But

$$E[Yg^T(X)] = E\{E[Y|X]g^T(X)\} = E[g(X)g^T(X)],$$

which implies

$$E\{[Y - g(X)][g(X) - \mu]^T\} = \phi.$$

Similarly it can be shown that

$$E\{[g(X) - \mu][Y - g(X)]^T\} = \phi.$$

Therefore,

$$V_Y = E\{[Y - g(X)][Y - g(X)]^T\} + V_{g(X)},$$

which implies that $V_Y - V_{g(X)}$ is positive semidefinite.

THEOREM 10-4.4. Let $X_1, X_2, \ldots, X_N$ be a random sample from a distribution that is a member of the family $\{f(x; \theta), \theta \in \Omega\}$. Let $Y_1 = u(X_1, X_2, \ldots, X_N)$ be a sufficient statistic for the parameter θ and let $Y_2(X_1, \ldots, X_N)$ be any unbiased statistic for the parameter θ. Let $E[Y_2 | y_1] = g(y_1)$. Then $V_{Y_2} - V_{g(Y_1)}$ is positive semidefinite.

Proof: The proof follows immediately from Theorem 10-3.3.

Theorem 10-3.4 implies that if one is given a sufficient and an unbiased statistic, Y_1 and Y_2, respectively, that a minimum variance unbiased statistic Y follows from the formula

$$Y = E[Y_2 | Y_1].$$

For a suggested reading list see References [87], [104], [105], and [158].

EXERCISES

1. Verify the covariance matrix R_V:

$$R_V = \begin{bmatrix} \Sigma & I \\ I & \Lambda \end{bmatrix}.$$

2. If

$$R_V = \begin{bmatrix} \Sigma & I \\ I & \Lambda \end{bmatrix} \quad \text{and} \quad R_V^{-1} = \begin{bmatrix} A_{11} & A_{12} \\ A_{21} & A_{22} \end{bmatrix},$$

then show that $A_{11}^{-1} = \Sigma - \Lambda$.

3. Let

$$Y = X\beta + U,$$

where X is a known $n \times p$ matrix such that $r(X) = q < \min(n, p)$, and U is an $n \times 1$ normal random vector whose mean vector is $E(U) = \Phi$ and has a known $n \times n$ covariance matrix

$$E(UU^T) = R.$$

What is the matrix lower bound for the covariance matrix of the vector estimator given by the Gauss–Markov theorem?

4. In Exercise 3, if $E(U) = \mu \neq 0$, what is the matrix lower bound?

5. Let $Y = Hx + U$, where H is an $n \times p$ matrix with all its columns linearly independent, x is a $p \times 1$ parameter vector, and U is an $n \times 1$ error vector distributed multivariate normal such that $E[U] = \phi$ and $E[U^T U] = \sigma^2 I$. Show that

$$\hat{X} = (H^T H)^{-1} H^T y \quad \text{and} \quad \hat{\sigma}^2 = \frac{Y^T[I - H(H^T H)^{-1} H^T]y}{n - p}$$

are sufficient estimators of x and σ^2, respectively.

6. Let Y be distributed bivariate normal; that is, let

$$f(y) = \frac{1}{(2\pi)|V|^{1/2}} e^{-(1/2)(y-\mu)^T V^{-1}(y-\mu)}, \quad -\infty < y < \infty,$$

where

$$V = \begin{bmatrix} \sigma_{11} & \sigma_{12} \\ \sigma_{21} & \sigma_{22} \end{bmatrix} \quad \text{and} \quad y - \mu = \begin{pmatrix} y_1 - \mu_1 \\ y_2 - \mu_2 \end{pmatrix}.$$

Let (y_{1i}, y_{2i}), $i = 1, 2, \ldots, n$ be a random sample; then show that

$$\hat{\mu}_1 = \frac{1}{n} \sum_{i=1}^{n} y_{1i},$$

$$\hat{\mu}_2 = \frac{1}{n} \sum_{i=1}^{n} y_{2i},$$

$$\hat{\sigma}_{11} = \frac{1}{n-1} \sum_{i=1}^{n} (y_{1i} - \bar{y}_1)^2,$$

$$\hat{\sigma}_{22} = \frac{1}{n-1} \sum_{i=1}^{n} (y_{2i} - \bar{y}_2)^2,$$

and

$$\hat{\sigma}_{12} = \frac{1}{n-1} \sum_{i=1}^{n} (y_{1i} - \bar{y}_1)(y_{2i} - \bar{y}_2)$$

are sufficient estimators of $\mu_1, \mu_2, \sigma_{12}, \sigma_{11}$, and σ_{22}, respectively.

CHAPTER 11

Least Square Estimates in a Nonlinear Model

11-1 Introduction

This chapter considers certain computational methods that may be employed in fitting nonlinear models by means of least squares and discusses methods and their advantageous properties for solving nonlinear least square problems. We first discuss a technique that incorporates Hartley's modified Gauss–Newton method and then Walling's iteration method, which makes valuable use of any linear parameters that may be present. A third technique, which incorporates the advantages of both previous techniques, is then discussed. Empirical results are included to indicate the number of iterations for convergence for each of the techniques.

11-2 Several Techniques for Solution of Nonlinear Least Square Problems

Given an n-dimensional input vector x with distinct elements x_j and an associated n-dimensional output vector y with elements y_j, we are frequently faced with finding some function of f of x and of an m-dimensional ($m \leq n$) parameter vector θ with elements θ_i such that f gives an approximate correspondence between x and y. Also, f is such that at least one of the m parameters is expressed nonlinearly in f. In most nonlinear least square problems, it is assumed that the mathematical form of the function f is known, and that we need only find the parameter vector θ that best describes the relationship $y_j = f(x_j; \theta), j = 1, 2, \ldots, n$, in the "least squares" sense. That is, we must determine the θ that minimizes

$$Q(x; \theta) = [y - f(x; \theta)]^T[y - f(x; \theta)], \qquad (11\text{-}1)$$

where $f(x; \theta)$ is the vector with elements $f(x_j; \theta), j = 1, 2, \ldots, n$.

The usual method employed in minimizing (11-1) is to select a suitable estimate θ_0 of θ and replace f in (11-1) by its approximate multiple first-order Taylor expansion about θ_0. If we define $f_i(x; \theta)$ to be the vector whose element are

$$\frac{\partial f(x_j; \theta)}{\partial \theta_i}, \qquad j = 1, 2, \ldots, n,$$

then we can let

$$Q_i(x; \theta) = \frac{\partial Q}{\partial \theta_i} = -2[y - f(x; \theta)]^T[f_i(x; \theta)]. \qquad (11\text{-}2)$$

The multiple first-order Taylor expansion of $f(x_j; \theta)$ about θ_0 can be written

$$p(x_j; \theta) = f(x_j; \theta_0) + \sum_{i=1}^{m} (\theta_i - \theta_{0i}) f_i(x_j; \theta_0) \qquad (11\text{-}3)$$

so that the problem becomes one of minimizing

$$Q(x; \theta) = [y - p(x; \theta)]^T[y - p(x; \theta)]. \qquad (11\text{-}4)$$

Setting the partial derivative of (11-4) with respect to each θ_i equal to zero, we obtain the following system of equations:

$$\begin{bmatrix} \sum f_1^2 & \sum f_1 f_2 & \cdots & \sum f_1 f_m \\ \sum f_1 f_2 & \sum f_2^2 & \cdots & \sum f_2 f_m \\ \vdots & \vdots & & \vdots \\ \sum f_1 f_m & \sum f_2 f_m & \cdots & \sum f_m^2 \end{bmatrix} \begin{bmatrix} d_1 \\ d_2 \\ \vdots \\ d_m \end{bmatrix} = -\frac{1}{2} \begin{bmatrix} Q_1 \\ Q_2 \\ \vdots \\ Q_m \end{bmatrix}, \qquad (11\text{-}5)$$

where $d_i = \theta_i - \theta_{0i}, i = 1, 2, \ldots, m$. Since the x_j's are distinct, the $(m \times m)$ matrix on the left of Equation (11-5) is positive definite, allowing one to solve for $d_i, i = 1, 2, \ldots, m$. Using the calculated d's one can obtain a new estimate $\theta_1 = \theta_0 + d$ of θ, where the elements of d are $d_i, i = 1, 2, \ldots, m$. Replacing θ_0 in (11-2) by θ_1 one can repeat the above process to obtain a new estimate θ_2 of θ.

Hartley's method is similar to the above with the addition of the following formal assumptions, which lead to the guaranteed convergence of the iterative procedure to the vector, which minimizes (11-1):

1. The first- and second-order partial derivatives of $f(x;\theta)$ with respect to the θ_i are assumed to exist and to be continuous for all vectors x.
2. It is assumed that for any nonzero vector u with elements u_i, $i = 1, 2, \ldots, n$ with $u^T u > 0$,

$$\sum_{j=1}^{n}\left[\sum_{i=1}^{m} u_i f_i(x_j;\theta)\right]^2 > 0 \qquad (11\text{-}6)$$

for every x and for all vectors θ in a bounded convex set S of the parameter space.
3. Denote by

$$Q = \lim_{\bar{S}} \inf\, Q(x;\theta),$$

where $\bar{S}$ is the complement of S given in assumption 2. Then it is assumed that it is possible to find a vector θ_0 in the interior of S such that

$$Q(x;\theta_0) < Q.$$

In addition, Hartley damps his vector d by a factor v, $0 \leq v \leq 1$, where v is chosen to minimize the function

$$Q(v) = Q(x;\theta_0 + vd). \qquad (11\text{-}7)$$

This number v is determined by noting that $Q(v)$ is, in the first-order Taylor expansion form of expression (11-4), approximately quadratic in v. Hence $Q(v)$ is evaluated at $v = 0$, $v = \frac{1}{2}$, and $v = 1$, and the parabola through the points obtained is seen to take a minimum at

$$v_{\min} = \frac{1}{2} + \frac{1}{4}\left[\frac{Q(0) - Q(1)}{Q(0) - 2Q\frac{1}{2} + Q(1)}\right].$$

The minimum $Q(v_{\min})$ should be compared to the value $Q(0)$. If the new value is not smaller, the computation should be done on a segment of half-length.

Having obtained the v with the desired properties, we form the new vector

$$\theta_1 = \theta_0 + vd$$

and proceed once more with the iteration.

The method developed by Walling differs from Hartley's method in that it makes use of the possible presence of one or more linear parameters in the function to which the given data are to be fitted. In addition, no original guesses need be made for the linear parameters, a fact that leads to simplified computation of the inverse of the ($m \times m$) matrix of Equation (11-5).

For the remainder of this chapter we assume that r and $m - r$ of the parameters in θ are linear and nonlinear, respectively, in f, where $0 < r < m$. We now make an initial guess for the nonlinear parameters, say, θ_{0i}, $i = r + 1, \ldots, m$. Then the problem is to fit the given "data" vectors to

$$Y \approx f[x; (\theta_1, \theta_2, \ldots, \theta_r, \theta_{0(r+1)}, \ldots, \theta_{0m})^T]. \qquad (11\text{-}9)$$

This is done by treating (11-9) in the usual linear least squares manner. Since we have a function of r linear parameters, setting

$$Q_i = \frac{\partial Q}{\partial \theta_i} = 0, \qquad i = 1, 2, \ldots, r \qquad (11\text{-}10)$$

we obtain r equations in r unknowns θ_i, $i = 1, 2, \ldots, r$. Solving this system of equations one obtains $\theta_0^* = (\theta_{01}, \theta_{02}, \ldots, \theta_{0r})^T$, which can be used to calculate more accurate estimates of the nonlinear parameters of f. This is accomplished by using the usual nonlinear least squares method, by forming Equation (11-5). From (11-10), we see that the first r of the elements in the right-hand vector of (11-5) are zero. In addition an ($r \times r$) block of the ($m \times m$) matrix of Equation (11-5) has already been calculated in determining the r linear parameters. We also have the inverse matrix of this block, which allows us to employ a process such as the bordering method to obtain the differences $d_{r+1}, d_{r+2}, \ldots, d_m$. These are then added to our initial guesses $\theta_{0(r+1)}, \ldots, \theta_{0m}$ to give us new estimates for the nonlinear parameters $\theta_{1(r+1)}, \ldots, \theta_{1m}$, which may then be used to obtain better choices for the linear parameters.

We now illustrate a method proposed by Nelson and Lewis, which takes full advantage of both the Hartley and Walling methods. Let us assume the assumptions 1, 2, and 3 (p. 171) are satisfied. We then apply Walling's method to obtain the $m - r$ differences $d_{r+1}, d_{r+2}, \ldots, d_m$. We now proceed as in Hartley's method to find the number v, $0 \leq v \leq 1$, which minimizes (11-7), where now $\theta_0 = (\theta_{01}, \theta_{02}, \ldots, \theta_{0r}, \theta_{0(r+1)}, \ldots, \theta_{0m})^T$ consists of $m - r$ initial guesses for the nonlinear parameters and r subsequent choices for the linear parameters. Also, the first r elements of the vector d are zero. We now calculate $Q(0)$, $Q(\frac{1}{2})$, and $Q(1)$ and find the number $v_{\min}$ at which the parabola passing through these values takes its minimum, again given by (11-8). Setting

$$\theta_{1i} = \theta_{0i} + v d_i, \qquad i = r + 1, \ldots, m \qquad (11\text{-}11)$$

11-3 A Numerical Example

Suppose we are given the following data:

$$x = \begin{bmatrix} -5 \\ -3 \\ -1 \\ 1 \\ 3 \\ 5 \end{bmatrix} \qquad y = \begin{bmatrix} 127 \\ 151 \\ 379 \\ 421 \\ 460 \\ 426 \end{bmatrix}$$

and we wish to fit the mathematical model

$$f(x;\theta) = \theta_1 + \theta_2 e^{\theta_3 x},$$

where θ_1 and θ_2 are linear parameters and θ_3 is a nonlinear parameter, to the above data.

Tables 1-8 present the first several cycles in each of the iterative procedures discussed. It should be noted that the values $Q(\theta)$ listed for Walling's method were evaluated after the calculation of a new nonlinear parameter, whereas Walling evaluates his $I(b) = [Q(\theta)]$ after the calculation of new linear parameters, thus accounting for the difference between the last column

Table 1. Iteration by the Usual Method with Starting Values (580, −180, −.16)

Cycle	θ_1	θ_2	θ_3	$Q(\theta)$
0	580.0000	−180.0000	−0.1600	27376.6184
1	490.4171	−121.1131	−0.2231	14585.8476
2	528.6932	−163.7856	−0.1851	13778.7834
3	515.9165	−148.5370	−0.2068	13407.5364
4	525.6375	−159.7482	−0.1965	13393.8133
5	522.0188	−155.4449	−0.2010	13390.5356
6	523.8065	−157.5399	−0.1991	13390.1696
7	523.0822	−156.6853	−0.1999	13390.1064
8	523.3989	−157.0579	−0.1996	13390.0953
9	523.2654	−156.9006	−0.1997	13390.0931

Table 2. Iteration by Hartley's Method with Starting Values (580, −180, −.16).

Cycle	θ_1	θ_2	θ_3	$Q(\theta)$
0	580.0000	−180.0000	−0.1600	27376.6184
1	495.2042	−124.2599	−0.2197	14590.5746
2	525.0784	−159.4128	−0.1905	13638.9330
3	519.3629	−152.4212	−0.2036	13394.5440
4	523.2578	−156.9350	−0.1995	13390.3250
5	523.2279	−156.8574	−0.1997	13390.0942
6	523.2911	−156.9313	−0.1997	13390.0930

Table 3. Iteration by Walling's Method with Starting values (580, −180, −.16).

Cycle	θ_1	θ_2	θ_3	$Q(\theta)$
0	580.0000	−180.0000	−0.1600	27376.6184
1	567.1459	−207.4688	−0.2148	38485.5644
2	510.8553	−142.1891	−0.1929	15566.6156
3	529.4858	−164.2012	−0.2025	13912.9616
4	520.8471	−154.0493	−0.1985	13473.8054
5	524.3828	−158.2155	−0.2002	13406.0975
6	522.8518	−156.4135	−0.1994	13392.9320
7	523.4994	−157.1760	−0.1998	13390.6102
8	523.2235	−156.8512	−0.1996	13390.1853
9	523.3405	−156.9890	−0.1997	13390.1088
10	523.2912	−156.9310	−0.1997	13390.0960
11	523.3117	−156.9551	−0.1997	13390.0935

Table 4. Iteration by the Proposed Method with Starting Values (580, −180, −.16).

Cycle	θ_1	θ_2	θ_3	$Q(\theta)$
0	580.0000	−180.0000	−0.1600	27376.6184
1	567.1459	−207.4688	−0.1758	15248.8192
2	547.3672	−184.9335	−0.1894	14592.2812
3	532.8699	−168.1533	−0.1984	13854.2130
4	524.4521	−158.2970	−0.1997	13398.9222
5	523.3086	−156.9515	−0.1997	13390.0931

in the representations of Walling's method that follow, and those previously published. It is felt that a more natural comparison between Walling's iteration and the other three methods results. In each table, the method used and the starting vector are stated.

Tables 1–8 show cycles for each iterative method for two different starting vectors θ_0. The method of Hartley is seen to converge to the desired parameter vector approximately three cycles faster than the usual method.

Sec. 11-3 A Numerical Example

Table 5. Iteration by the Usual Method with Starting Values (500, −140, −.18).

Cycle	θ_1	θ_2	θ_3	$Q(\theta)$
0	500.0000	−140.0000	−0.1800	18282.5078
1	512.3155	−144.5639	−0.2153	13504.4028
2	527.5772	−162.2170	−0.1931	13412.2712
3	520.4815	−153.6758	−0.2025	13392.3969
4	524.3173	−158.1482	−0.1984	13390.4604
5	522.8298	−156.3895	−0.2002	13390.1540
6	523.5007	−157.1780	−0.1994	13390.1035
7	523.2210	−156.8483	−0.1998	13390.0949
8	523.3413	−156.9900	−0.1996	13390.0932

Table 6. Iteration by Hartley's Method with Starting Values (500, −140, −.18).

Cycle	θ_1	θ_2	θ_3	$Q(\theta)$
0	500.0000	−140.0000	−0.1800	18282.5078
1	511.3856	−144.2193	−0.2127	13435.7050
2	521.2696	−155.1402	−0.2008	13392.1877
3	523.5613	−157.2929	−0.1993	13390.1408
4	523.2576	−156.9024	−0.1997	13390.0942
5	523.3046	−156.9489	−0.1997	13390.0927

Table 7. Iteration by Walling's Method with Starting Values (500, −140, −.18).

Cycle	θ_1	θ_2	θ_3	$Q(\theta)$
0	500.0000	−140.0000	−0.1800	18282.5078
1	542.6366	−179.4838	−0.2076	18417.4696
2	516.5619	−148.9786	−0.1962	14023.6732
3	526.4170	−160.6055	−0.2011	13523.1848
4	522.0257	−155.4399	−0.1990	13412.7574
5	523.8593	−157.5997	−0.1999	13394.3216
6	523.0713	−156.6720	−0.1996	13390.8518
7	523.4057	−157.0659	−0.1997	13390.2310
8	523.2631	−156.8979	−0.1996	13390.1177
9	523.3238	−156.9693	−0.1997	13390.0974
10	523.2976	−156.9385	−0.1997	13390.0936

The method of Walling does not appear to converge significantly faster than the usual method, but it should be pointed out that the initial guesses for the linear parameters are not used in the iterative process, and thus do not contribute to a more rapid convergence. As already stated, other factors render this method more advantageous in many cases.

Table 8. Iteration by the Proposed Method with Starting Values (500, −140, −.18).

Cycle	θ_1	θ_2	θ_3	$Q(\theta)$
0	500.0000	−140.0000	−0.1800	18282.5078
1	542.6366	−179.4838	−0.1929	14419.2484
2	529.5462	−164.2720	−0.1993	13621.4810
3	523.6539	−157.3580	−0.1997	13390.9023
4	523.3079	−156.9506	−0.1997	13390.0931

The proposed method is seen to converge at least as fast as, and for many cycles faster than, any of the other three methods. This fact was further borne out in subsequent iterations using several other starting values. In no case was the proposed method found to be inferior to any of the other three discussed after five iterative cycles.

There have been many computing algorithms developed for computing estimates of nonlinear estimates. The interested reader should consult the References.

EXERCISES

1. Perform all the necessary mathematical steps to obtain Equation (11-5).

2. By assuming the x's to be distinct, prove that the matrix on the left of Equation (11-5) is positive definite.

3. Verify that (11-6) is another way of saying that

$$\begin{bmatrix} \sum_{i=1}^{n} f_1(x_1;\theta) & \sum_{i=1}^{n} f_1(x_1;\theta)f_2(x_2;\theta) & \cdots & \sum_{i=1}^{n} f_1(x_i;\theta)f_n(x_i;\theta) \\ \vdots & \vdots & & \vdots \\ \sum_{i=1}^{n} f_m(x_i;\theta)f_1(x_i;\theta) & \sum_{i=1}^{n} f_m(x_i;\theta)f_2(x_2;\theta) & \cdots & \sum_{i=1}^{n} f_m^2(x_i;\theta) \end{bmatrix}$$

is positive definite.

4. Verify Equation (11-8).

REFERENCES

1. Aitken, A. C., "On Least Squares and Linear Combinations of Observations," *Proc. Roy. Soc. Edinburgh, A,* **55**, 1935, pp. 42–48.
2. Aitken, A. C., "Studies in Practical Mathematics II. The Evaluation of the Latent Roots and Latent Vectors of a Matrix," Proc. Roy. Soc. Edinburgh, A, **62**, 1948, pp. 273–277.
3. Aitken, A. C., "On the Statistical Independence of Quadratic Forms in Normal Variates," Biometrika, **37**, 1950, pp. 93–96.
4. Albert, A., "Real Time Computation of Constrained Least Squares Estimators," *ARCON—Advanced Research Consultants,* Lexington, Mass., June 1965.
5. Albert, A., and Sittler, R. W., "A Method for Computing Least Squares Estimators That Keep Up With the Data," *J. Soc. Ind. Appl. Math. Ser. A,* 1965, pp. 384–417.
6. Anderson, C. L., "A Geometric Theory of Pseudoinverses and Some Applications in Statistics," Master's Thesis, Southern Methodist University, Dallas, Tex., 1967.
7. Anderson, T. W., "The Integral of a Symmetric Unimodal Function Over a Symmetric Convex Set and Some Probability Inequalities," *Proc. Amer. Math. Soc.,* **6**, 1955, pp. 170–176.
8. Anderson, T. W., *An Introduction to Multi-Variate Statistical Analysis,* New York: Wiley, 1958.
9. Anderson, T. W., "Least Square and Best Unbiased Estimates," *Ann. Math. Statist.* **33**: 1, March 1962, p. 272.
10. Ashar, V., and Wallace, T. D., "A Sampling Study of Minimum Absolute Deviations Estimator," *Operations Res.,* **11**, Sept.–Oct. 1963, pp. 747–758.
11. Balakrishnan, A. V., "On a Characterization of Processes for Which Optimal Mean-Square Systems Are of a Specified Form," *IRE, Trans. PGIT,* **6**, Sept. 1960, pp. 491–500.
12. Balakrishnan, A. V., "Report on Progress in Information Theory in the U.S.A., 1960–1963: Prediction and Filtering," *IEEE, Trans. PGIT,* **IT-9**, Oct. 1963, pp. 237–239.
13. Battin, R. H., "A Statistical Optimization Navigation Procedure for Space Flight," *A.R.S. J.,* **32**, 1962, pp. 1681–1696.
14. Bauer, F. L., "Elimination with Weighted Row Combinations for Solving Linear Equations and Least Squares Problems," *Numer. Math.,* **VII**, 1965, pp. 338–352.

15. Bendat, J. S., "A General Theory of Linear Prediction and Filtering," *J. Soc. Ind. Appl. Math.*, **4**, Sept. 1956, pp. 131–151.

16. Bendat, J. S., "Exact Integral Equation Solutions and Synthesis for a Large Class of Optimum Time Variable Linear Filters," *IRE, Trans. PGIT*, **IT-3**, March 1957, pp. 71–80.

17. Bendat, J. S. and Piersol, A. G., *Measurement and Analysis of Random Data*, New York: Wiley, 1966.

18. Benedict, T. R., and Sondhi, M. M., "On a Property of Wiener Filters," *Proc. IRE*, **45**, July 1957, pp. 1021–1022.

19. Ben-Israel, A., and Dohen, D., "On Iterative Computation of Generalized Inverses and Associated Projections," *J. Soc. Ind. Appl. Math. Numer. Anal.*, **III**, 1966, pp. 410–419.

20. Ben-Israel, A., and Wersan, S. J., "An Elimination Method for Computing the Generalized Inverse of an Arbitrary Complex Matrix," *J. Assoc. Comput. Mach.*, **X**, 1963, pp. 532–537.

21. Berkson, J., "Estimation by Least Squares and by Maximum Likelihood," *Proceedings of the Third Berkeley Symposium on Mathematical Statistics and Probability*, Vol. 1, Berkeley, Calif.: Univ. of California Press, 1956, pp. 1–11.

22. Blum, M., "Fixed Memory Least Squares Filters Using Recursive Methods," *IRE Trans. Inform. Theory*, **IT-3**: 3, Sept. 1957.

23. Blum, M., "On Exponential Filters," *J. Assoc. Comput. Mach.*, **6**: 2, April, 1959, pp. 283–304.

24. Blum, M., "A Stagewise Parameter Estimation Procedure for Correlated Data," *Numerische Mathematik*, **3**, 1961, pp. 202–208.

25. Blum, M., "Best Linear Unbiased Estimation by Recursive Methods," *J. Soc. Ind. Appl. Math.*, **14**: 1, 1966, pp. 167–180.

26. Bode, H. W., and Shannon, C. E., "A Simplified Derivation of Linear Least-Squares Smoothing and Prediction Theory," *Proc. IRE*, **38**, April 1950, pp. 417–425.

27. Boot, J. C. G., *Quadratic Programming*, Amsterdam: North Holland; Skokie, Ill.: Rand McNally, 1964.

28. Booton, R. C., Jr., "An Optimization Theory for Time-Varying Linear Systems with Non-stationary Statistical Inputs," *Proc. IRE*, **40**, Aug. 1952, pp. 977–981.

29. Born, H. G., and Tapley, B. D., "Sequential Estimation of State and the Observation Error Covariance Matrix," *Amer. Instit. Aero. Astro.*, 1969.

30. Bose, R. C., *Lecture Notes on Analysis of Variance*, Chapel Hill, N.C.: The Univ. of North Carolina Press, 1959.

31. Brown, J. L., Jr., "Asymmetric Non-Mean-Square Error Criteria," *IRE, Trans. PGAC*, **AC-7**, Jan. 1962, pp. 64–66.

32. Bryson, A. E., Jr., and Johansen, D. E., "Linear Filtering for Time-Varying Systems Using Measurements Containing Colored Noise," *Research Report 385*, Jan. 1964, Applied Research Lab., Sylvania Electronic Systems, 40 Sylvan Rd., Waltham, Mass.

33. Cheney, E. W., *Introduction to Approximation Theory*, New York: McGraw-Hill, 1966.

34. Chipman, J. S., "On Least Squares with Insufficient Observations," *J. Amer. Statist. Assoc.*, **59**, 1964, pp. 1078–1111.

35. Chipman, J. S., and Rao, M. M., "Projections, Generalized Inverses, and Quadratic Forms," *J. Math. Anal. Appl.*, **IX**, 1964, pp. 1–11.

36. Chipman, J. S., and Rao M.M., "On the Treatment of Linear Restrictions in Regression Analysis," *Econometrica*, **XXXII**, 1964, pp. 198–209.

37. Claus, A. J., R. B. Blackman, E. G. Halline, and W. C. Ridgway, III, "Orbit Determination and Prediction, and Computer Programs," *Bell System Tech. J.*, **42**, July 1963, pp. 1357–1382.

38. Cline, R. E., "Note on the Generalized Inverse of the Product of Matrices," *SIAM Rev.*, **6**, Jan. 1964, pp. 57–58.

39. Cox, H., "On the Estimation of State Variables and Parameters for Noise Dynamic Systems," *IEEE, Trans. PGAC*, **AC-9**, Jan. 1964, pp. 5–12.

40. Cramer, H., "A Contribution to the Theory of Statistical Estimation," *Skand. Akwarietidskrift*, **29**, 1946, p. 85.

41. Cramer, H., *Mathematical Methods of Statistics*, Princeton, N.J.: Princeton Univ. Press, 1946.

42. Dantzig, G. B., and Orden, A., "Duality Theorems," *RAND Report R. M.-1265*, The RAND Corp., Santa Monica, Calif., Oct. 1953.

43. Darlington, S., "Linear Least-Squares Smoothing and Prediction, with Applications," *Bell System Tech. J.*, **37**, Sept. 1958, pp. 1221–1293.

44. Darlington, S., "Nonstationary Smoothing and Prediction Using Network Theory Concepts," *IRE, Trans. PGIT*, **IT-5**, Special Supplement, May 1959, pp. 1–13.

45. David, F. N., and Johnson, N. L., "Statistical Treatment of Censored Data, Part I, Fundamental Formula," *Biometrika*, **41**, 1954, pp. 228–240.

46. David, F. N., and Neyman, J., "Extension of the Markoff Theorem on Least Squares, *Statist. Res. Mem.*, **2**, 1938, pp. 105–116.

47. Decell, H. P., Jr., "An Application of Generalized Matrix Inversion to Sequential Least Squares Parameter Estimation," *NASA TN D-2830*. May, 1965.

48. Decell, H. P., Jr., and Odell, P. L., "A Note on a Generalization of the Gauss–Markov Theorem," *Texas J. Sci.*, March 1966, pp. 21–24.

49. DeGroot, M. H., and Rao, M. M., "Bayes Estimation with Convex Loss," *Ann. Math. Statist.*, **34**, Sept. 1963, pp. 839–846.

50. Deutsch, R., *Estimation Theory*, Englewood Cliffs, N.J.: Prentice-Hall, 1965.

51. Dorn, W. S., "Duality in Quadratic Programming," *Quart. Appl. Math.* **18**, 1960, pp. 155–162.

52. Durbin, J., "A Note on Regression When There Is Extraneous Information About One of the Coefficients," *J. Amer. Statist. Assoc.*, **48**, Dec. 1953, pp. 799–808.

53. Dwyer, P. S., "Generalizations of a Gaussian Theorem," *Ann. Math. Statist.*, **22**: 1, 1958, pp. 106–117.

54. Dwyer, P. S., and Macphail, M.S., "Symbolic Matrix Derivatives," *Ann. Math. Statist.*, **XIX**, 1948, pp. 517–534.

55. Fisher, R. A., "The Goodness of Fit and Regression Formulae, and the Distribution of Regression Coefficients," *J. Roy. Statist. Soc.*, **85**, Part IV, 1922, reprinted in *Contributions to Mathematical Statistics*, New York: Wiley, 1950.

56. Fisher, W. D., "A Note on Curve Fitting with Minimum Deviations by Linear Programming," *J. Amer. Statist. Assoc.*, **56**, 1961, pp. 359–362.

57. Forsythe, G. E., "Theory of Selected Methods of Finite Matrix Inversion and Decomposition," *Nat. Bur. of Std. (U.S.)*, *INA-52-5*, August 1951.

58. Foster, M., "An Application of the Wiener–Kolmogorov Smoothing Theory to Matrix Inversion," *J. Soc. Ind. Appl. Math.*, **9**, 3, Sept. 1961, 387–392.

59. Frame, J. S., "Part I—Matrix Functions and Applications," *IEEE Spectrum*, Mar. 1964, pp. 208–220.

60. Franklin, Joel N., "Numerical Simulation of Stationary and Nonstationary Gaussian Random Processes," *SIAM Rev.*, **7**: 1, Jan. 1965.

61. Friedland, B., "Least Squares Filtering and Prediction of Nonstationary Sampled Data," *Information and Control*, **1**, 1958, pp. 297–313.

62. Friedman, B., *Principles and Techniques of Applied Mathematics*, New York: Wiley, 1957,

63. Gainer, P. A., "A Method for Computing the Effect of an Additional Observation on a Previous Least-Square Estimate," *NASA TN D-1599*, 1964.

64. Gantmacher, F. R., *The Theory Matrices*, Vols. 1 and 2, translated by K. A. Hirsch, New York: Chelsea, 1959.

65. Gately, W. Y., "Application of the Generalized Inverse Concept to the Theory of Linear Statistical Models," Doctoral Dissertation, Oklahoma State University, Stillwater, Okla., 1962.

66. Gauss, C. F., "Theoria Combinationis Observatorium, Para Posterior (1821), Werke, 4-31; Supplement (1826), Werke 4:71.

67. Glicksman, A. M., *An Introduction to Linear Programming and the Theory of Games*, New York: Wiley, 1963.

68. Goldberger, A. S., "Note on Stepwise Least Squares," *J. Amer. Statist. Assoc.*, **56**, 1961, pp. 105–110.

69. Goldebrger, A. S., "Stepwise Least Squares: Residual Analysis and Specification Error," *J. Amer. Statist. Assoc.*, **56**, 1961, pp. 998–1000.

70. Goldberger, A S., *Econometric Theory*, New York: Wiley, 1964.

71. Goldman, A. J., and Zelen, M., "Weak Generalized Inverse and Minimum Variance Linear Unbiased Estimation," *J. Res. Nat. Bur. Std., B Mathematics and Mathematical Physics*, **68B**, 4, 1964, pp. 151–172.

72. Golub, G. H., "Comparison of the Variance of Minimum Variance and Weighted Least Squares Regression Coefficients," *Ann. Math, Statist.*, **34**, Sept. 1963, pp. 984–991.

73. Golub, G. H., "Numerical Methods for Solving Linear Least Squares Proglems," *Numer. Math.*, **VII**, 1965, pp. 206–216.

74. Golub, G. H., and Kahan, W., "Calculating the Singular Values and Pseudo-inverse of a Matrix," *J. Soc. Ind. Appl. Math, Numer. Anal.*, **II**, 1965, pp. 205–224.

75. Goodman, L. A., "A Further Note on Miller's Finite Markov Processes in Psychology," *Psychometrika*, **18**, 1953, pp. 245–248.

76. Graybill, F. A., *An Introduction to Linear Statistical Models*, **1**, New York: McGraw-Hill, 1961.

77. Graybill, F. A., and Deal, R. B., "Combining Unbiased Estimators," *Biometrics*, **15**, 1959, pp. 543–550.

78. Grenander, U., and Rosenblatt, M., *Statistical Analysis of Stationary Time Series*, New York: Wiley, 1950.

79. Greville, T. N. E., "The Pseudoinverse of a Rectangular or Singular Matrix and Its Application to the Solution of Systems of Linear Equations," *SIAM Rev.*, **1**, 1959, pp. 38–43.
80. Greville, T. N. E., "Note on the Generalized Inverse of a Matrix Product," *SIAM Rev.*, **8**, 1966, pp. 518–524.
81. Haavelmo, T., "The Statistical Implications of a System of Simultaneous Equations," *Econometrica*, **11**, 1943, pp. 1–12.
82. Halmos, P. R., *Finite-Dimensional Vector Spaces*, 2nd ed., New York: Van Nostrand Reinhold, 1958.
83. Halperin, Max, "Normal Regression Theory in the Presence of Intraclass Correlation," *Ann. Math, Statist.* **22**, 1951, pp. 575–580.
84. Hartley, H. O., "The Modified Gauss–Newton Method for the Fitting of Non-linear Regression Functions by Least Squares," *Technometrics*, **III**: 2, 1961, pp. 269–280.
85. Ho, Y. C., "On the Stochastic Approximation Method and Optimal Filtering Theory," *J. Math. Anal. Appl.*, **6**, 1962, pp. 152–154.
86. Ho, Y. C., and Lee, R. C. K., "A Bayesian Approach to Problems in Stochastic Estimation and Control," *IEEE, Trans. PGAC*, **AC-9**, Oct. 1964, pp. 333–339.
87. Hogg, H. V., and Craig, A. T., *Introduction to Mathematical Statistics*, 2nd ed., New York: Macmillan, 1965.
88. Hohn, F. E., *Elementary Matrix Algebra*, 2nd ed., New York: Macmillan, 1964.
89. Householder, A. S., *Principles of Numerical Analysis*, New York: McGraw-Hill, 1953.
90. Householder, A. S., "A Class of Methods for Inverting Matrices," *J. Soc. Ind. Appl. Math.*, **6**, 2, June 1958, pp. 189–195.
91. Householder, A. S., *The Theory of Matrices for Numerical Analysis*, Waltham, Mass.: Blaisdell, 1964.
92. Huzurbazar, V. S., "The Likelihood Equation, Consistency and the Maxima of the Likelihood Function," *Ann. Eugen., Lond.*, **14**, 1948, pp. 185–200.
93. Johnston, J. *Econometric Methods*, New York: McGraw-Hill, 1963.
94. Judge, G. G., and Takayama, T., "Inequality Restrictions in Regression Analysis," *J. Amer. Statist. Assoc.*, **61**, 1966, pp. 166–81.
95. Kale, B. K., "On the Solution of the Likelihood Equation by Iteration Processes," *Biometrika*, **48**, 1961, pp. 452–456.

96. Kalman, R. E., "Contributions to the Theory of Optimal Control," *Bol. Soc. Mat. Mexicana*, V, 1960, pp. 102–119.
97. Kalman, R. E., "A New Approach to Linear Filtering and Prediction Problems," *J. Basic Eng.*, Trans. ASME, **82D**, 1960, pp. 33–45.
98. Kalman, R. E., "New Methods and Results in Linear Prediction and Filtering Theory," *Tech. Rept. 61–1*, Research Institute for Advanced Studies, Baltimore, Md., Mov. *1960*.
99. Kalman, R. E., "On the General Theory of Control Systems," *Proceedings of the First Congress of the International Federation of Automatic Control, Moscow*, 1960, London: Butterworth, 1961.
100. Kalman, R. E., and Bucy, R. S., "New Results in Linear Filtering and Prediction Theory," *J. Basic Eng.*, Trans. ASME, **83D**, 1961, pp. 95–108.
101. Karlin, S., *Mathematical Methods and Theory in Games, Programming and Economics*, London: Addison-Wesley, 1959.
102. Karts, O. J., "Linear Curve Fitting Using Least Deviations," *J. Amer. Statist Assoc.*, **53**, March 1958, pp. 118–132.
103. Keller, H. B., "On the Solution of Singular and Semidefinite Linear Systems by Iteration," *J. SIAM Numer. Anal.*, **II**, 1965, pp. 281–290.
104. Kendall, M. G., and Stuart, S., *The Advanced Theory of Statistics*, Vol. 2, London: Griffin, 1961.
105. Kendall, M.G., and Stuart, S., *The Advanced Theory of Statistics*, Vol. 1, 2nd ed. New York: Hafner, 1963.
106. Kiefer, J., and Wolfowitz, J., "Stochastic Estimation of the Maximum of a Regression Function," *Ann Math. Statist.*, **23**, July 1952, pp. 462–466.
107. Kolmogoroff, A., "Interpolation and Extrapolation Von Stationaren Zufalligen Folgen," *Bull. Acad. Sci. U.S.S.E., Sr. Math.* 1941, **5**, pp. 3–14.
108. Korganoff, A., "The Inversion of Rectangular Matrices in the Resolution of Ill-Conditioned Linear Systems," *Proc. Nordsam Congress*, Helsinki, **2**, (Aug. 16–20 1963); Helsinki, 1964, **2**, pp. 179–190.
109. Kruskal, W, "The Coordinate-free Approach to Gauss–Markov Estimation and Its Application to Missing and Extra Observations," *Proceedings of the Fourth Berkeley Symposium on Mathematical Statistics and Probablity* Vol. 1, Berkeley, Calif: Univ. of California Press, pp. 435–451.
110. Kruskal, W, "When Are Gauss–Markov and Least Squares Estimators Identical? A Coordinate-free Approach," *Ann. Math. Statist.*, **39**: 1, 1968, pp. 70–75.

111. Kuhn, H., and Tucker, A., "Non-linear Programming," *Proceedings of the Second Berkeley Symposium*, H. Neyman, Ed., Berkeley, Calif.: Univ. of California Press, 1951, pp. 481–492.

112. Lancaster, P., *Theory of Matrices*, New York and London: Academic Press, 1969.

113. Lee, T. C., Judge, G. G., and Takayama, T., "On Estimating the Transitional Probabilities of a Markov Process, *J. Farm Econ.* **47**, 1965, pp. 742–762.

114. Lehman, E. K., *Notes on the Theory of Estimation*, Berkeley, Calif.: Univ. of California Press, 1962.

115. Levenberg, K., "A Method for the Solution of Certain Non-linear Problems in Least Squares," *Quart. Appl. Math.*, **2**, 1944, pp. 164–168.

116. Levine, N., "A New Technique for Increasing the Flexibility of Recursive Least Squares Data Smoothing," *Bell System Tech. J.*, **40**, May 1961, pp. 821–840.

117. Lewis, T. O., "Application of the Theory of Generalized Matrix Inversion to Statistics," Doctoral Dissertation, Univ. of Texas, 1966.

118. Lewis, T. O. and Newman, T., "Pseudoinverses of Positive Semi-definite Matrices," *J. Soc. Ind. Appl. Math.*, pp. 701–703, July 1968, Vol. 16, No. 4.

119. Lewis, T. O., and Odell, P. L., "A Generalization of Gauss–Markov Theorem," *J. Amer. Statist. Assoc.*, **61**, 1966, pp. 1063–1066.

120. Lindgren, B. W., *Statistical Theory*, New York: Macmillan, 1962.

121. Linnik, Yu. V., *Method of Least Squares and Principles of the Theory of Observation*, London: Pergamon Press, 1961.

122. Loève, M., *Probability Theory*, 3rd ed., New York: Van Nostrand Reinhold, 1963.

123. McElroy, F. W., "A Necessary and Sufficient Condition That Ordinary Least-Squares Estimators Be Best Linear Unbiased," *J. Amer. Statist. Assoc.*, **62**, Dec. 1967, pp. 1302–1304.

124. Madansky, A., "Least Squares Estimation in Finite Markov Processes," *Psychometrika*, **24**, 1959, pp. 137–144.

125. Magness, T. A., and McGuire, J. B., "Comparison of Least Squares and Minimum Variance Estimates of Regression Parameters," *Ann. Math. Statist.*, **33**, 1962, pp. 462–470.

126. Marcus, M., "Basic Theorems in Matrix Theory," *Nat. Bur. Std. (U.S.) Appl. Math. Ser.* 15, Jan. 22, 1960.

127. Maxfield, J. E., and Gardner, R. S., "Note on Linear Hypotheses with Prescribed Matrix of Normal Equations," *Ann. Math. Statist.*, **26**, 1955, pp. 149–150.

128. Mayne, D. Q., "Optimal Non-stationary Estimation of the Parameters of a Linear System with Gaussian Inputs," *J. Electron. Control*, Jan. 1963, pp. 101–112.

129. Meyer, J., and Glauber, R. R., *Investment Decisions, Economic Forecasting and Public Policy*, Boston Division of Research, Graduate School of Business Administration, Harvard Univ. 1964.

130. Middleton, D., *An Introduction to Statistical Communication Theory*, New York: McGraw-Hill, 1960.

131. Miller, G. A., "Finite Markov Processes in Psychology," *Psychometrika*, **17**, 1952, pp. 149–167.

132. Mitra, S. K., "On a Generalized Inverse of a Matrix and Applications," *Sankhya*, Series A, **XXX**: 1, 1968, pp. 107–114.

133. Mood, A. M., and Graybill, F. A., *Introduction to the Theory of Statistics*, 2nd ed., New York: McGraw-Hill, 1963.

134. Moore, E. H., Abstract, *Bull. Amer. Math. Soc.*, **26**, 1920, pp. 394–395.

135. Morris, G. L., and Odell, P. L., "A Characterization for Generalized Inverses of Matrices," *SIAM Rev.*, 1968, pp. 208–211.

136. Morris, G. L., and Odell, P. L., "Common Solutions for n Matrix Equations with Applications," *J. Assoc. Comput. Mach.*, **15**, 1968, pp. 272–274.

137. Murphy, G. J., and Sahara, K., "A Mean-Weighted Square-Error Criterion for Optimum Filtering of Nonstationary Random Processes," *IRE, Trans. PGAC*, **AC-6**, May 1961, pp. 211–216.

138. Nelson, D. L., "Numerical Methods for the Solution of Non-Linear Least Squares Problems," Master's Thesis, Texas Tech. University, Lubbock; Tex., 1969.

139. Nelson, D. L., "Quadratic Programming Techniques Using Matrix Pseudoinverses," Doctoral Dissertation, Texas Tech. University, Lubbock, Tex, 1969.

140. Osborne, E. E., "On Least Squares Solutions of Linear Equations," *J. Assoc. Comput, Mach.*, **8**, 1961, pp. 628–636.

141. Osborne, E. E., "Smallest Least Squares Solutions of Linear Equations," *J. Soc. Ind. Appl. Math. Numer. Anal.*, **II**, 1965, pp. 300–307.

142. Penrose, R. A., "A Generalized Inverse for Matrices," *Proc. Cambridge Phil. Soc.*, **51**, 1955, pp. 406–413.

143. Penrose, R. A., "On Best Approximate Solutions of Linear Matrix Equations," *Proc. Cambridge Phil. Soc.*, **LII**, 1956, pp. 17–19.

144. Perlis, S., *Theory of Matrices*, Reading, Mass.: Addison-Wesley, 1958.

145. Plackett, R. L., "Some Theorems in Least Squares," *Biometrika*, **37**, 1950, pp. 149–157.

146. Plackett, R. L., *Principles of Regression Analysis*, Fair Lawn, N.J.: Oxford, 1960.

147. Price, C. M., "The Matrix Pseudoinverse and Minimal Variance Estimates," *SIAM Rev.*, **6**, 22, 1964, pp. 115–120.

148. Pugachev, V. S., "General Condition for the Minimum Mean Square Error in a Dynamic System," *Avt. i Telemekh*, **17**, 1956, pp. 289–295, translation, pp. 307–314.

149. Pugachev, V. S., "The Method of Determining Optimum Systems Using General Bayes Criteria," *IRE, Trans. PGCT*, **CT-7**, Dec. 1960, pp. 491–505.

150. Pugachev, V. S., "Statistical Theory of Systems Reducible to Linear," *IRE Trans. PGCT*, **CT-7**, Dec. 1960, pp. 506–513.

151. Raiffia, A., and Schlaifer, R., *Applied Statistical Decision Theory*, Boston: Harvard Univ. Press, 1961.

152. Rainbolt, M. B., "Sequential Least Squares Parameter Estimation," *NASA-MSC* unpublished report, 1964.

153. Rao, C. R., "Generalization of Markoff's Theorem and Tests of Linear Hypotheses," *Sankhya*, **7**, 1945–1946, pp. 9–19.

154. Rao, C. R., "On the Linear Combination of Observations and the General Theory of Least Squares," *Sankhya*, **7**, 1945–1946, pp. 237–256.

155. Rao, C. R., "A Theorem in Least Squares," *Sankhya*, **11**, 1951, pp. 9–12.

156. Rao, C. R., "On Transformations Useful in the Distribution Problems of Least Squares," *Sankhya*, **12**, 1952–1953, pp. 339–346.

157. Rao, C. R., "A Note on a Generalized Inverse of a Matrix with Applications to Problems in Mathematical Statistics," *J. Roy. Statist. Soc., Ser. B*, **24**, 1962, pp. 152–158.

158. Rao, C. R., *Linear Statistic Inference and Its Applications*, New York: Wiley, 1966.

159. Rao, C. R., "Generalized Inverse for Matrices and Its Applications in Mathematical Statistics," *Festschrift for J. Neyman: Research Papers in Statistics*, London: Wiley, 1966, pp. 263–279.

160. Rao, C. R., "Least Squares Theory Using an Estimated Dispersion Matrix and Its Application to Measurement of Signals," *Proceedings of the Fifth Berkeley Symposium on Mathematical Statistics and Probability*, Vol. 1, Berkeley Calif., University of Calif Press. 1967, pp. 355–371.

161. Rao, K. K., "A Simplified Proof of Gauss–Markov Theorem When the Regression Matrix Is of Less Than Full Rank," *Amer. Math. Monthly*, **LXXIII**, 1966, pp. 394–395.

162. Rice, J. R., *The Approximation of Functions*, Vol. 1, *Linear Theory*, Reading, Mass.: Addison-Wesley, 1964.

163. Rice, J. R., and White, J. S., "Norms for Smoothing and Estimation," *SIAM Rev.*, **6**: 3, July 1964, pp. 243–256.

164. Rohde, C. A., "Contributions to the Theory, Computation and Applications of Generalized Inverses," Doctoral Dissertation, North Carolina State University, Raleigh, N.C., 1964.

165. Rohde, C. A., "Some Results on Generalized Inverses," *SIAM Rev.*, **VIII**: 2, 1966, pp. 201–205.

166. Rohde, C. A., and Harvey, J. R., "Unified Least Squares Analysis," *J. Amer. Statist. Assoc.* **LX**, 1965, pp. 523–527.

167. Saaty, T. L., and Bram, J., *Non-linear Mathematics*, New York: McGraw-Hill, 1964.

168. Searle, S. R., "Additional Results Concerning Estimable Functions and Generalized Inverse Matrices," *J. Roy. Statist. Soc. Ser. B*, **XXVII**, 1965, pp. 486–490.

169. Searle, S. R., *Matrix Algebra for the Biological Sciences*, New York: Wiley, 1966.

170. Sherman, S., "A Theorem on Convex Sets with Applications," *Ann. Math. Statist.*, **26**, 1955, 763–767.

171. Sherman, S., "Non-Mean-Square Error Criteria," *IRE, Trans. PGIT*, **IT-4**, Sept. 1958, pp. 125–126.

172. Shumway, R. H., and Dean, W., "Best Linear Unbiased Estimation for Multivariate Stationary Processes," *Technometrics*, **10**, 1968, pp. 523–534.

173. Singleton, R. R., "A Method for Minimizing the Sum of Absolute Values of Deviations," *Ann. Math, Statist.*, **11**, 1940, p. 301.

174. Slepian, D., "Estimation of Signal Parameters in the Presence of Noise," *IRE, Trans. PGIT*, **IT-3**, March 1964, pp. 68–89.

175. Smith, G. L., Schmidt, S. F., and McGee, L. A., "Application of Statistical Filter Theory to the Optimal Estimation of Position and

Velocity on Board a Circumlunar Vehicle," *Tech. Rept.* R-135, Nov. 20, 1961, Ames Research Center, NASA, Moffett Field, Calif.

176. Swerling, P., "A Proposed Stagewise Differential Correction Procedure for Satellite Tracking and Prediction," *Report* P-1292, Jan. 8, 1958, The Rand Corp., Santa Monica, Calif.

177. Swerling, P., "First-Order Error Propagation in a Stagewise Smoothing Procedure for Satellite Observations," *J. Astronaut. Sci.*, **6**, 1959, pp. 46–52.

178. Swerling, P., "Parameter Estimation Accuracy Formulas," *IEEE Trans. PGIT*, **IT-10**, Oct. 1964, pp. 302–314.

179. Takayama, T., and Judge, G. G., "Equilibrium Among Spatially Separated Markets: a Reformulation," *Econometrica*, **32**, 1964, pp. 510–524.

180. Tapley, B. D., and Odell, P. L., *Texas Center for Research Quarterly Progress Report No. 1*, Austin, Tex., June 1964.

181. Telser, L. G., "Least Squares Estimates of Transition Probabilities," *Measurement in Economics*, Stanford, Calif.: Stanford Univ. Press, 1963, pp. 272–292.

182. Theil, H., *Economic Forecasts and Policy*, 2nd ed., Amsterdam: North Holland, 1961.

183. Theil, H., "On the Use of Incomplete Prior Information in Regression Analysis," *J. Amer. Statist. Assoc.*, **58**, June 1963, pp. 401–414.

184. Theil, H., and Goldberger, A. S., "On pure and Mixed Statistical Estimation in Economics," *Intern. Econ. Rev.* **2**, Jan. 1961, pp. 65–78.

185. Theil, H., and Rey, G., "A Quadratic Programming Approach to Estimation of Transition Probabilities," *Management Sci.* **12**: 1966, pp. 714–721.

186. Their, H., and van de Panne, C., "Quadratic Programming as an Extension of Classical Quadratic Maximization," *Management Sci.*, 7: 1, 1960, pp. 1–20.

187. Tiano, G. C., and Zellner, A., "Bayes Theorem and the Use of Prior Knowledge in Regression Analysis," *Biometrika*, **51**: 1 and 2, 1964, pp. 219–230.

188. Tintner, G., *Econometrics*, New York: Wiley, 1952.

189. Tintner, G., "Stochastic Linear Programming with Applications to Agricultural Economics," *Proceedings of the Second Symposium in Linear Programming*, Washington, D.C., 197–228.

190. Tucker, H. G., *A Graduate Course in Probability*, New York and London: Academic Press, 1967.

191. Wagner, H. M., "Linear Programming for Regression Analysis," *J. Amer. Statist. Assoc.*, **54**, 1959, pp. 206–212.

192. Walling, D. D., "Non-linear Least Squares Curve Fitting When Some Parameters Are Linear." *Texas J. Sci.*, **20**: 2, Dec. 10, 1968, pp. 119–124.

193. Watson, G. S., "Linear Least Squares Regression," *Ann. Math. Statist.*, **38**, 1961, pp. 1679–1699.

194. Watson, G. S., "Serial Correlation in Regression Analysis I," *Biometrika*, **42**, Dec. 1955, pp. 327–431.

195. Whittle, P., *Prediction and Regulation by Linear Least-Square Methods*, London: The English Universities Press, Ltd., 1963.

196. Wiener, N., *The Extrapolation, Interpolation and Smoothing of Stationary Time Series*, New York: Wiley, 1949.

197. Wilde, D. J., and Beightler, C. S., *Foundations of Optimization*, Englewood Cliffs, N.J.: Prentice-Hall, 1967.

198. Wilks, S. S., *Mathematical Statistics*, Princeton, N.J.: Princeton Univ. Press, 1945.

199. Wolfe, P., "The Simplex Method for Quadratic Programming," *Econometrica*, **27**, 1959, pp. 382–398.

200. Wong. E., and Thomas, J. B., "On the Multidimensional Prediction and Filtering Problem and the Factorization of Special Matrices," *J. Franklin Inst.*, **272**, Aug. 1961, pp. 87–99.

201. Zadeh, L. A., and Desorer, C. A., *Linear System Theory*, New York: McGraw-Hill, 1963.

202. Zadeh, L. A., and Ragazzini, J. R., "An Extension of Wiener's Theory of Prediction," *J. Appl. Phys.*, **21**, July 1950, pp. 645–655.

203. Zadeh, L. A., and Ragazzini, J. R., "Optimum Filters for the Detection of Signals in Noise," *Proc. IRE*, **40**, Oct. 1952, pp. 1223–1231.

204. Zakai, M., "On a Property of Wiener Filters," *IRE Trans, PGIT*, **IT-5**, March 1959, pp. 15–17.

205. Zelen, M., and Godman, A. J., "Weak Generalized Inverses and Minimum Variance Linear Unbiased Estimation," *Tech. Report* **314**, Mathematics Research Center, U. S. Army, University of Wisconsin, Madison, 1963.

206. Zellner, A., "Linear Regression with Inequality Constraints on the Coefficients," *Mimeographed Report 6109* of the *International Center for Management Science*, 1961.

207. Zellner, A., "An Efficient Method of Estimating Seemingly Unrelated Regressions and Tests for Aggregation Bias," *J. Amer. Statist. Assoc.*, **57**, 1862, pp. 348–368.

208. Zellner, A., and Theil, H., "Three-Stage Least Squares: Simultaneous Estimation of Simultaneous Equations," *Econometrica*, **30**, 1962, pp. 54–78.

209. Zyskind, George, "On Canonical Forms, Nonnegative Covariance Matrices and Best and Simple Least Squares Linear Estimators in Linear Models," *Ann. Math, Statist.*, **38**, 1967, pp. 1092–1109.

Index

A

A posteriori risk, 43

B

Best approximate solution, 9–10
    unique best approximate solution, 9

C

Characteristic function, 33
Characteristic root or Eigenvalue, 13
Cofactor, 12
Complement, 4
Complex Euclidean n-space, 2
Conditional expectation, 43
Correlation, 120
    autocorrelation, 120
    cross correlation, 120
Covariance, 30
    autocovariance or covariance kernel, 120
    covariance adjustment, 139–140
    cross covariance, 120
Crout factorization, 16

D

Density function, 25–27
    bivariate normal, 27, 44, 162

Density function (*cont.*):
    conditional, 26
    continuous, 25, 27
    marginal, 29
    normal, 27–29, 31–35
Derivatives of determinants and matrices, 17–18
    determinants, 17
    equation, 18
    matrices, 18
Dirac delta functional, 90–91, 120–121
Distribution function, 25–26
    absolutely continuous, 25
    discrete, 25
Dynamic model, 119–133

E

Estimators, 35–44
    bayesian estimator, 42, 49
    best, 37
    best linear estimator, 48
    best linear recursive estimators, 75–84
    best linear unbiased estimators, 55, 86–90
    BLUE (best linear unbiased estimator), 55
    combining estimators and observations, 69–71
    consistent, 38

Index

Estimators (*cont.*):
  estimating subvectors, 66
  inequality constraints, 100, 103–117
  least squares, 59–60, 135
  linear constraints with additive random components, 102
  linear estimators with linear constraints, 99
  maximum likelihood, 38
  mean-squared error, 37, 67
  minimized mean-squared error, 37, 48, 67–68
  minimum variance, 37, 59–60, 100
  optimal, 127, 133
  recursive, 60–62, 71, 77
  sufficiency, 161–166
  unbiased, 36–37
  using continuous data, 85–98

F

Filter functions, 124
Fourier transform, 34, 86–90

G

Gauss-Markov theorem, 52–58
  extension, 56–58
  random parameter vector, 62–65

H

Hartley's method, 169, 171, 174–175

I

Inside-Out rule, 16

K

Kalman-Bucy filter, 133
Kronecker delta operator, 87
Kuhn-Tucker theorem, 106

L

Lagrange multiplier matrix, 19–20, 65, 101
Linear equations, 5
  best approximate solution, 9
  general solution of a system of linear equations, 5, 10
  unique best approximate solution, 9
Linear model, 50, 99
  relations between discrete and continuous models, 94–96
Loss function, 42
  matrix valued squared-error loss function, 43–44

M

Matrices, 3–12
  brace, 3
  column space, 4
  equivalence, 6
  fundamental, 130
  generalized matrix inverse, 6
  hermition, 7, 12–15
  identity, 5
  inverse, 5
  left inverse, 5
  norm, 4, 8
  normalized generalized inverse, 6
  non-negative definite, 127
  non-singular, 12
  null, 4
  partitioned, 13
  positive definite, 13–15, 18–19, 22
  positive semidefinite, 13–15, 18–19, 22
  product, 11
  pseudo inverse, 6
  range space, 4
  rank, 5
  reflexive generalized inverse, 6
  right inverse, 5
  spectra, 88
  transition, 130–131
Matrix lower bound, 157–160

Matrix variational, 18, 19
Minor, 17

N

Nelson and Lewis method, 172–174, 176

O

Optimal design matrix, 145–155
Orthogonal, 3
    spaces, 3, 4
    vectors, 3
Orthogonality principle, 67–69
Orthogonal projection theorem, 122
    Wiener-Hopf equation, 123

P

Pitman, 45
Pseudo inverse of a matrix produce, 11

Q

Quadratice forms, 12
    positive definite, 13
    positive semi-definite, 13

R

Random variables, 23–46
    continuous, 25
    covariance, 30
    discrete, 25
    expected value, 30
    moments, 30
    variance, 30
Random vectors, 24

S

Sample space, 21
    elementary event, 21

Sample space (*cont.*):
    event, 21
    probability function, 21
    sigma algebra, 21
    sure event, 21
Selection of sample points, 145–155
    rectangular and ellipsoidal regions, 151–154
    rotation, 152
    translation, 152
Spectra matrix, 88
State vector, 50
Statistical independence, 27
Statistic, 36, 123
    estimators, 36, 37
Stochastic process, 119–120
Sufficient estimators, 161–166
    factorization theorems, 163–164
    Rao-Blackwell theorem, 165

T

Tableau, 108–111

V

Vector space, 1
Vectors, 1
    associative law, 1
    commutative law, 1
    distributive law, 1
    inner product, 2
    linear independence, 2
    multiplication of vectors by scalars, 1
    norm, 4
    transpose, 1
    vector addition, 1

W

Walling's method, 171–175
Wiener-Hopf equation, 123
Wolfe algorithm, 103, 107

TS
9/7/71